T0310282

Explosion Systems with Inert High-Modulus Components

Increasing the Efficiency of Blast Technologies and Their Applications

Igor A. Balagansky, Anatoliy A. Bataev, and Ivan A. Bataev

This edition first published 2019

Registered Office
John Wiley & Sons, Inc., 111 River Street, Hoboken, NJ 07030, USA

Editorial Office
111 River Street, Hoboken, NJ 07030, USA

For details of our global editorial offices, customer services, and more information about Wiley products visit us at www.wiley.com.

Wiley also publishes its books in a variety of electronic formats and by print-on-demand. Some content that appears in standard print versions of this book may not be available in other formats.

Library of Congress Cataloging-in-Publication Data

Names: Balagansky, I. A. (Igor Andreevich), 1952– author. | Bataev, A. A. (Anatoliy), author. | Bataev, I. A. (Ivan). author.
Title: Explosion systems with inert high modulus components : increasing the efficiency of blast technologies and their applications / I.A. Balagansky, A.A. Bataev, I.A. Bataev.
Description: First edition. | Hoboken, NJ : Wiley, 2019. | Includes bibliographical references and index. |
Identifiers: LCCN 2019011946 (print) | LCCN 2019017020 (ebook) | ISBN 9781119525424 (Adobe PDF) | ISBN 9781119525394 (ePub) | ISBN 9781119525448 (hardback)
Subjects: LCSH: Detonation waves. | Explosions. | Blast effect. | Fracture mechanics. | Shaped charges.
Classification: LCC QC168 (ebook) | LCC QC168 .B335 2019 (print) | DDC 662.2–dc23
LC record available at https://lccn.loc.gov/2019011946

Cover design: Wiley
Cover image: Courtesy of Igor A. Balagansky, Anatoliy A. Bataev, Ivan A. Bataev

Set in 10/12pt Warnock by SPi Global, Pondicherry, India

Printed in the United States of America

V10010259_051419

Contents

Preface

Physical and physicochemical processes that are described by the concept "explosion" are characterized by release of a large amount of energy in a confined space within short periods of time. Not going deeply into the description of the variety of such processes, which were considered in detail in widely known monographs, we would like to point out that many features of explosive processes still remain the subject of thorough studies. While initially the phenomenon of explosion was studied almost exclusively for subsequent military applications, nowadays the received knowledge has found wide technological application. For example, explosion for ejection is widely used during construction of canals and mining of pits. One of the most striking examples of a directional explosion for ejection is the constructing of an anti-mudflow dam in the Medeo Valley (the Almaty Region, Kazakhstan) in 1968.

Out of many applications of explosion, the following examples may be particularly emphasized as most widely used in industry:

- Explosive forming – fabrication of tank bottoms, expansion of pipes, extrusion, punching and final calibration of pre-prepared products.
- Explosive welding – cladding of materials for chemical industry, welding of pipes, fabrication of composite materials, welding of materials, which cannot be joined using common technologies of fusion welding (i.e. Ti-Fe, Al-Cu, Al-Fe, etc.).
- Explosive cutting – separation of materials, parts, or constructions when common technologies of cutting are inapplicable. The typical examples are separation of rocket stages, dismantling of thick-walled autoclaves at nuclear power plants, and disposing of ships, tanks, etc.
- Explosive hardening – hardening of railway frogs, excavator bucket teeth, parts of mining machines, internal surfaces of various mixers, mills, etc.
- Explosive compaction – fabrication of billets from powder and porous materials.

- Shock-assisted synthesis of materials – fabrication of diamond from carbon, synthesis of cubic boron nitride, synthesis of novel metastable materials.
- Synthesis of rapidly quenched materials – fabrication of amorphous and micro- and nanocrystalline alloys with specific properties.
- Modification of properties of high temperature superconductors aiming to increase the superconducting transition temperature.

Among other applications of explosion, one may mention mining, perforation of oil wells, generation of mega-gauss magnetic fields and mega-ampere current pulses, acceleration of bodies to hypersound velocities, etc.

Military applications of explosion do also continue to develop and improve. In this field it is necessary to mention the improvement in the methods of explosion destruction of fortifications, fragmentation of shells, formation of cumulative jets for damaging tanks armor, initiation of nuclear reactions by transition of a nuclear charge to a supercritical state, etc.

The detonation process is of particular importance to understand the phenomena occurring in chemical explosives. According to one of the definitions, the detonation is a stationary process of propagation of a narrow zone of exothermic chemical reactions excited by a shock wave compressing an explosive, which in turn is maintained by the released energy. A complex of a shock wave and a zone of exothermic chemical reactions are called a detonation wave. When developing explosive devices and technologies, the corresponding estimates and calculations are most often based on the hydrodynamic theory (model) of detonation developed in the works of Zeldovich–Neumann–Döring. The initial assumptions of this model are formulated for the case of a plane detonation wave, which obviously does not correspond to the detonation of real charges, which always have finite dimensions.

This leads to concepts of limiting and critical diameters (thicknesses) of charges. However, a discussion of the degree of adequacy and the limits of applicability of the hydrodynamic model continues in the scientific community. A number of effects were found in multiple experiments that do not fit into its framework. For example, there is a discussion of data on chemical reactions behind the Chapman–Jouguet plane (i.e. behind the postulated reaction zone). In experiments with charges of complex geometric shapes, in experiments where the detonation front comes out into the expanding region, or in the presence of cavities or gaps between charges and inert elements, the disturbances in the stationarity of the process are detected (zones of unreacted explosive appear). Particular attention should be paid to detonation processes in charges bounding with inert elements of materials having a sound velocity exceeding detonation velocity. In such cases, shock waves generated by detonation in inert materials can outpace the front of the detonation wave and compress the explosive substance ahead of the front. This leads to a change in the state of explosives and a corresponding change in the kinetics of the detonation transformation. As a result, the stationarity of the detonation process is disrupted, which can lead to changes in its course that are difficult to predict. From a practical point of view, this may cause a decrease or an increase in the effectiveness of explosive technologies. On the other hand, an understanding of the mechanisms of generation, development, and realization of nonstationary effects in detonation can provide a novel approach toward increasing the effectiveness of explosive technologies and/or to purposeful control of detonation processes.

In this monograph we will attempt to systematically describe available data obtained in the experimental study of nonstationary detonation processes in explosive charges containing inert elements made of materials which sound velocity significantly exceeds the detonation velocity. The experimental data is analyzed and compared with simulation results of the corresponding problems in an attempt to explain the physical mechanisms of the observed effects.

The first chapter contains examples of nonstationary propagation of detonation in real situations in the presence of wave perturbations ahead of the detonation front.

The second chapter presents the results of experimental studies of detonation processes in charges containing ceramic rods. The effects of the shock front blurring, the desensitization of explosives under the emerging compression wave, and the possibility of Mach configurations forming on the detonation front are discussed. The experimental results are confirmed by corresponding numerical calculations.

In the third chapter we discuss the results of experimental studies and numerical simulations of nonstationary processes on the contact surface between explosives and inert plates. The data obtained indicates that the detonation near the explosive/ceramic interface has a highly nonstationary nature. This is manifested in inconstancy of pressure values and in differences between the detonation velocity and the stationary detonation parameters. We show that the perturbations arising at the interface affect the front of the detonation wave and change its shape.

The fourth chapter is devoted to the study of peculiarities of detonation processes in cylindrical explosive charges placed in inert shells. Comparing the explosions occurring in copper and ceramic shells, we show both experimentally and by numerical simulations that despite practically the same dynamic rigidity of the shell materials, there are significant differences in wave processes both in detonation products and in shells. This occurs due to the differences in sound velocities between copper and silicon carbide and due to the rapid destruction of the ceramics under explosive loading. A mechanism for transferring perturbations from the periphery to the symmetry axis of cylindrical explosive charges is discussed.

In the fifth chapter, cumulative processes arising in the collapse of cylindrical shells are analyzed. When ceramic tubes are compressed, a stream of brightly glowing particles is formed, which under certain conditions can be called a cumulative jet. The leading part of this stream propagates at a speed approximately two times the detonation velocity and contains the high speed and main parts. This stream of particles displays initiating and penetrative capabilities when acting upon explosives.

The sixth chapter discusses the possibility of application of monolithic and disperse ceramic materials in specific structures that protect dangerous substances during their transportation and storage.

The seventh chapter describes the structures of metals that appear under loading using explosive systems with high-modulus ceramic elements. The possibility to calculate the parameters of explosive loading by observing the resulting material structures is also considered.

At the end of the monograph, a list of notable publications by the authors on the topic under consideration is presented.

Appendix A contains the information on mechanical properties and behavior peculiarities of a number of high-modulus materials under impact loading. Special attention is given to the properties of ceramic materials, which were used in the research by the book authors.

Appendix B describes the experimental methods and simulation software, which were used to prepare this book.

This monograph does not pretend to provide exhaustive coverage of the problems mentioned in it. It was prepared based on results of the research carried out by the authors at the Novosibrisk State Technical University in collaboration with scientific groups of the Institute of Problems of Chemical Physics of the Russian Academy of Sciences (Chernogolovka, Moscow Region, Russia), the Bauman Moscow State Technical University (Moscow, Russia), the Lavrentyev Institute of Hydrodynamics of the Siberian Branch of the Russian Academy of Sciences (Novosibirsk, Russia), and the Institute of Pulsed Power Science, Kumamoto University (Kumamoto, Japan). The main results presented in this book were obtained with help of L.A. Merzhievsky, A.V. Utkin, S.V. Razorenov, I.F. Kobylkin, E.F. Gryaznov, A.D. Matrosov, I.A. Stadnichenko, K. Hokamoto, and P. Manikandan. Our pleasant duty is to thank all the colleagues who took part in these studies. Without their participation and consultations, this work would have never been completed.

1

Examples of Nonstationary Propagation of Detonation in Real Processes

Nonstationary propagation of detonation is frequently observed in charges encapsulated in shells or located in blastholes and boreholes when explosives are used for crushing rocks or intensifying oil extraction. It is also observed in charges with elongated cavities. This effect is related to the appearance and propagation of wave disturbances and jets moving ahead of the detonation front. This can lead both to the interruption of detonation and to its accelerated propagation. Under certain conditions, a pulsating behavior can occur when the detonation velocity is changing periodically. This chapter discusses some examples of these processes.

1.1 Channel Effect

The channel effect is the phenomenon first described by T. Urbański [1, 2], which manifests itself in the fact that in the presence of an air gap between the wall of the shell or the hole and the explosive charge, the charge detonates with a variable velocity, changing as it propagates. Depending on the properties of a specific explosive, it is possible either to increase the propagation velocity or to decrease it up to the complete interruption of the detonation. Under certain conditions, a pulsating detonation behavior may be observed. An increase in the detonation velocity is observed in the case of sensitive powdered explosives like hexogen (RDX); a decrease is typical for ammonites and other mixed explosives. The pulsating behavior is observed in plastic explosives of the dynamite type.

L.V. Dubnov and L.D. Khotina [3], L.V. Dubnov et al. [4], A.I. Golbinder and V.F. Tyshkevich [5, 6], and A.N. Dremin et al. [7] have shown that the channel effect depends on the properties of explosives and it also

Explosion Systems with Inert High-Modulus Components: Increasing the Efficiency of Blast Technologies and Their Applications, First Edition. Igor A. Balagansky, Anatoliy A. Bataev, and Ivan A. Bataev.

increases with increasing mass and strength of the shell and the length of the charge. High roughness of the walls of the shell and the charge surfaces, as well as filling the gap with paper and cardboard baffle, reduce the channel effect. A powerful additional detonator, in contrast to blasting in open air, accelerates the damping of detonation in ammonite-like explosives. The smaller the diameter of the charge, the more likely the detonation damping may occur due to the channel effect.

Sufficiently complete description of the mechanisms leading to the channel effect may be found in [3]. This paper describes the experiments carried out using a high-speed camera and pulsed X-ray radiography. The photograph in Figure 1.1 shows the processes occurring under the influence of the channel effect during explosion of ammonite PZhV-20. One may clearly observe a luminescence zone in the explosive corresponding to the detonation wave and a glowing zone in the channel that gradually expands. This indicates that the interface between the glowing medium in the channel and the unperturbed air moves in the channel at a faster rate than the detonation velocity in the explosive. The scheme of the explosive charge deformation due to the channel effect is shown in Figure 1.2.

The following qualities of detonation under the influence of the channel effect were revealed:

- During explosion of ammonite PZhV-20, the detonation velocity decreases continuously, down to complete attenuation. The velocity of the glowing wave in the gap also gradually decreases, but slower. As a result, the difference between the velocity of the wave in the gap ω and the velocity of the detonation front D increases with time.

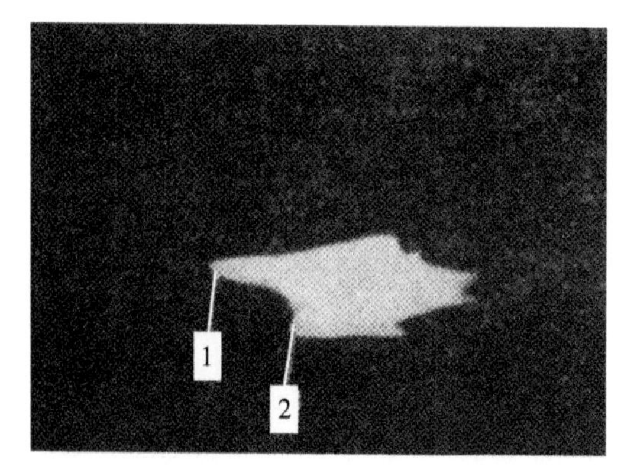

Figure 1.1 Photograph of the channel effect: 1, front of the shock wave in the gap and 2, detonation front. (*Source:* From Dubnov and Khotina [3]. Reprinted with permission of Springer Nature.)

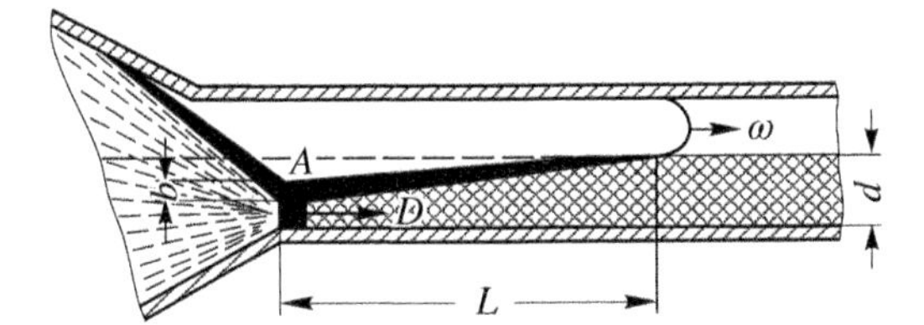

Figure 1.2 Scheme of deformation of the explosive charge under the channel effect: ω is the air wave front in the gap; D is the detonation velocity; d is the initial diameter of the explosive charge; b is the thickness of the compressed zone; and L is the length of the compressed zone. (*Source:* From Dubnov and Khotina [3]. Reprinted with permission of Springer Nature.)

- At the beginning of explosion of the powdered trinitrotoluene (TNT), both speeds initially increase, but the growth rate of D is greater than that of ω. At a certain distance from the initiation point, both velocities become equal and the entire complex moves further with an almost constant speed that exceeds the detonation velocity of a similar charge of TNT exploded in the open air. The final velocity in this case corresponds to the detonation velocity of TNT with a density of around 1.3 g/cm^3.
- For phlegmatized powdered RDX, the detonation proceeds in a similar manner, but with a slightly shorter velocity alignment period. The stabilized detonation velocity is 6.83 km/s, which corresponds to the detonation velocity of the RDX with a density of 1.23 g/cm^3.
- For pressed TNT, the distance and the period of the velocity alignment is less than for powdered one. The increase in the detonation velocity is approximately 400 m/s.

The studies of the channel effect performed using X-ray imaging have shown that the material ahead of the detonation front is densified and the charge cross section is reduced. One may conclude that an air shock wave propagates through the gap, and in all of the studied cases at the initial period its velocity is $\omega > D$. After that, depending on the effect of densification on the detonation properties of explosives, the velocities can be equalized.

As a result of a channel shock wave formation, the density of the charge ahead of the detonation front increases. The changes in the area of the charge cross section and formation of a zone of increased density can in turn affect the detonation process. The charge ahead of the detonation front can also be densified by a shock wave propagating inside the shell.

In the studies simulating the channel effect in boreholes, it was established that it appears when the size of the gap between the charge and the wall of the shell is in the range of 0.1–3 charge diameters and reaches the highest value at some optimal gap, which depends on specific conditions of the process. Further studies of the channel effect have allowed to obtain detailed information of the mechanisms of its development. Due to the rapid expansion of practical applications of emulsion explosives, the channel effect was recently studied for these materials as well [8].

1.2 Detonation of Elongated High Explosive Charges with Cavities

Nonstationary detonation processes, which are caused by specific channel effects, may be observed in elongated explosive charges with cavities. Apparently, for the first time, such processes were described in studies [9–12]. It is well known that due to the cumulative phenomena inside the cavity of the detonating tubular charge, a powerful shock wave appears. Over a length of several tens of the cavity diameter, the velocity of the wave is almost constant and higher than the normal detonation velocity (~7.5% higher for dense explosives and ~65% higher for low-density substances).

The following scheme of the process is proposed based on experimental data. As mentioned earlier, a powerful shock wave appears inside the cavity of the charge due to the cumulative phenomena. The velocity of this wave is constant over the length of several tens of the charge diameters. After that, the shock wave decays due to losses of interaction with the walls of the charge. During the decay of the primary shock wave in the region between its front and the detonation front, a secondary shock wave is generated, the velocity of which also significantly exceeds the detonation velocity. In this case, the velocity of detonation propagation along the outer surface of the charge in a small initial region is equal to the usual detonation velocity for a given explosive, and then it abruptly increases up to some value that weakly depends on the initial density of explosive. The magnitude of the jump is a few percent for dense explosives and a few dozen percent for low-density explosives.

The schematic description of the detonation process in such a structure is shown in Figure 1.3 [13], where *QR* is an unperturbed section of the initial plane detonation wave, *SS'* is a front of an air shock wave in the channel. This shock wave creates a compression wave *SP* inside the charge. Inside the region, *OSP* explosive is compressed and its detonation velocity exceeds the initial one so that the detonation wave *OP* propagates ahead of the unperturbed wave *QR*. The part of the detonation wave *PQ* moves with the initial velocity along the uncompressed explosive. At some moment in time, the point *Q* reaches the outer surface of

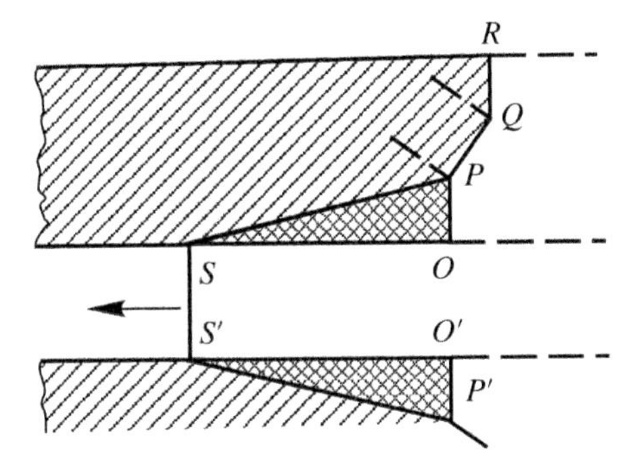

Figure 1.3 Structure of the process of detonation propagation. (*Source:* From Zagumennov et al. [13]. Reprinted with permission of Springer Nature.)

the charge, which manifests itself as a sudden increase in the detonation velocity observed on the surface.

The experiments carried out with the cavities having large elongation (i.e. large ratio of the length of the channel to its diameter) have shown that in such cases, as a rule, there is a transient process that ends up with the establishment of a stationary regime at a sufficiently large distance from the beginning of the cavity. In this stationary regime, the channel wave moves consistently ahead of the detonation front. In the transient mode, two channel waves are observed. In this case, the secondary wave velocity oscillates due to the fact that initially it propagates along the region of the nonuniform flow created by the primary wave. As follows from [13], with a further increase in elongation a third wave can appear, and so on. This iterative formation of compression waves can be explained by the fact that when the shock wave propagates in the cavity, the length of the column of compressed gas increases as the process spreads [14]. At the same time, its energy decreases due to the radial expansion, so the velocity of the leading front of the wave and the mass velocity behind its front will also decrease. When the gas velocity near the detonation front becomes lower than the detonation velocity of explosives, the products of the explosion will affect the gas again and a new shock wave will be created in it. The second shock wave catches up with the first one and passes through it. If the length of the charge is sufficiently large, this effect can be repeated periodically.

If the channel terminates inside the charge, then, providing sufficient sensitivity of explosive, a wave propagating inside the channel can initiate the detonation. In this case, two detonation waves will propagate from the point of initiation in opposite directions, one of which will move backward and meet the initial detonation wave.

In charges prepared from explosives, the detonation ability (i.e. sensitivity) of which decreases with increasing density, the presence of a channel may cause an attenuation of detonation.

The increase in detonation velocity observed in tubular charges can be used to reduce the time of detonation transfer in charges of a given length and to increase their brisance. Local areas of increased brisance can be created by using short annular sections, which provide initiation of detonation at the closed end of the channel.

In study [15], it was theoretically shown that in charges of sufficiently sensitive explosives having a longitudinal channel (or several channels), a self-sustaining "superfast" detonation process is possible, the velocity of which is several times greater than the usually observed minimum Chapman–Jouguet detonation velocity for that substance. This conclusion does not contradict the known fundamental assumptions of the detonation theory, which in principle allows any detonation velocity greater than the minimum one, providing there exists a mechanism that ensures the initiation of a heat release reaction with this velocity. In this case, the initiation is provided by a shock wave in the gas that fills the channel. The efficient initiation of explosive surrounding the channel in a super-high-velocity regime requires an increased initial gas density in comparison with atmospheric air and a sufficiently large channel size.

The phenomenon of detonation in charges with longitudinal channels has been simulated multiple times for a wide variety of initial setups. One of the examples is a simple one-dimensional model that reproduces the main peculiarities of the detonation process, in which detonation is treated according to the classical Chapman–Jouguet and Zeldovich-Neumann-Döring approaches. This model may be found in [16]. In [17], a numerical two-dimensional model of the detonation of a porous cylindrical charge of pentaerythritol tetranitrate (PETN) with a cavity is described. One of the conclusions of the simulation was that the compression caused by a wave propagating in the channel initiates an advanced detonation. The results of the calculations were in a good agreement with experimental observations and with the theoretical model described in [15].

Thus, one may observe two possible routes of evolution of accelerated detonation propagation in charges with longitudinal cavities.

1) A powerful shock wave and a jet of detonation products occur in the cavity due to cumulative effects and outrun the detonation front, thus compressing the adjacent layer of unreacted explosives. In this case, the detonation wave in this layer propagates along the explosive with a density greater than the initial one [13].
2) The initiation of explosives occurs from the cavity surfaces at a higher rate than the detonation velocity.

The phenomenon of accelerated propagation of detonation under certain conditions is observed when the cavity is filled with an inert

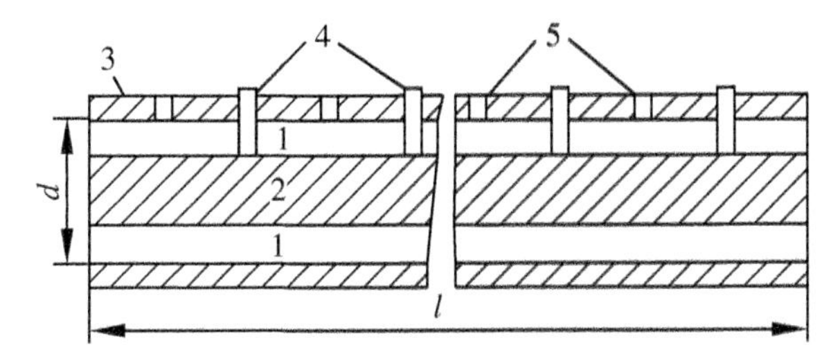

Figure 1.4 The scheme of the experimental assembly: 1, explosive charge; 2, lithium rod; 3, shell; 4, optical fibers; and 5, holes in the shell. (*Source:* From Merzhievskii et al. [18]. Reprinted with permission of Springer Nature.)

material. This process has been experimentally studied in [18, 19]. These papers describe experimental studies of detonation in cylindrical charges, the axial channel of which was filled with lithium. The powdered RDX with the initial density $\rho_0 \sim 0.85–1.1\,\text{g/cm}^3$ and PETN ($\rho_0 \sim 0.9\,\text{g/cm}^3$) were used as explosives. The scheme of experimental setup is shown in Figure 1.4. The internal diameter of the channel was 10 mm, the outer diameter of the charge d varied from 30 to 60 mm, and the length l was in range from 25 to 300 mm. The charges were mounted on a thin paper or aluminum alloy shell. The output of the detonation front and the shock wave propagating in lithium on the butt surface of the charge or propagation of the detonation along the charge were studied using a high-speed camera. In the latter case, the glowing of the detonation wave on the surface of the charge and from the interface layer between the charge and the lithium was recorded using special optical fibers made of glass capillaries with a diameter of 2 mm.

Figure 1.5 shows a typical photographic record of output of the detonation front and shock wave in lithium on the end surface of the charge (RDX, $\rho_0 \approx 1.1\,\text{g/cm}^3$, $d = 40\,\text{mm}$, $l = 55\,\text{mm}$). Figure 1.6 presents the photograph of the process propagation along the charge of RDX ($\rho_0 \approx 0.85\,\text{g/cm}^3$, $d = 36\,\text{mm}$, $l = 300\,\text{mm}$).

The experimental studies provide clear evidence that the detonation front is inclined toward the axis of the charge. The propagation velocity of the process in the direction of the charge axis D_1 exceeds the normal detonation velocity D for the explosive of a given density. For RDX ($\rho_0 = 1.1\,\text{g/cm}^3$, $d = 40\,\text{mm}$, $l = 55\,\text{mm}$), the excess of D_1 over D is approximately $22.5 \pm 2.5\%$. The inclination angle of the detonation front β initially increases, but then, after the distance of approximately 4–5 diameters of the internal channel, the process reaches a stationary mode and β remains unchanged.

The articles give an approximate estimate of the maximum possible velocity of the process propagation, which can be achieved by the

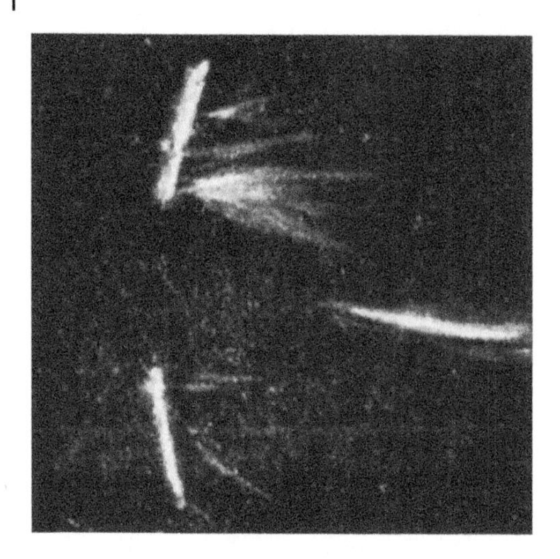

Figure 1.5 Typical photographic record of output of the detonation front and shock wave in lithium on the end surface of the charge. (*Source:* From Merzhievskii et al. [18]. Reprinted with permission of Springer Nature.)

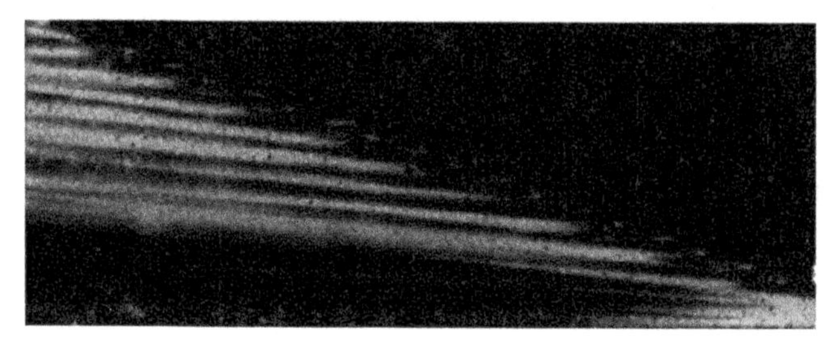

Figure 1.6 Photograph of the process propagation along the charge of hexogen (RDX). (*Source:* From Merzhievskii et al. [18]. Reprinted with permission of Springer Nature.)

proposed method in combinations of explosive and filler. For RDX with $\rho_0 = 1.1\,\text{g/cm}^3$, $D_{max} = 7.7\,\text{km/s}$ and the value of D_1 obtained in the experiments was 7.6 km/s, that is, the maximum rate was almost reached. In order to obtain higher speeds, it is recommended to use more powerful and sensitive explosives in combination with a less dense filler. The analysis carried out in [18, 19] has shown that accelerated propagation of reaction was due to the initiation of detonation from the surface of a cavity filled with lithium.

1.3 The Effects of Wall and Shell Material, Having Sound Velocity Greater Than Detonation Velocity, on the Detonation Process

Another class of nonstationary detonation propagation processes may be observed in experiments where the charge is in contact with shells or walls made of materials in which the sound velocity exceeds the detonation velocity. An example of such situation is described in [20], where the effect of an inert beryllium wall on the propagation of detonation in a charge of a low-sensitive triaminotrinitrobenzene (TATB)-based explosive EDC35 was investigated. It was observed that when the detonation wave propagates along a wall of an inert material, the velocity of the shock wave excited in the wall is usually lower than the detonation velocity. Thus, the shock wave propagates behind the detonation front at an angle to the surface equal to $\sin^{-1}(U/D)$, where U is the velocity of the shock wave and D is the detonation velocity. However, by carefully selecting a combination of the explosive and the material of the inert wall, it is possible to observe a situation when the velocity of the shock wave exceeds the detonation velocity. Currently there are only few confirmations that the leading wave in the inert wall can affect the local detonation velocity. The experiments described in [20] had the intention to show the degree of this effect on the detonation velocity in EDC35.

Figure 1.7 shows the experimental setup, which includes (A) a flat charge of EDC35 explosive with thickness of 25 mm, placed between (B) 9.3 mm thick beryllium plate and (C) 10 mm thick bronze plate. The charge was initiated by a 76.2 mm line initiator, with a 12.7 mm square cross-section Composition B/perspex smoothing system (D), followed by a 25 mm square cross-section Composition B booster (E). Coaxial pin probes used to measure the detonation velocity were placed along the center line in the bronze plate in contact with the explosive charge. The main experimental technique was a video recording using a high-speed slit-type camera with a rotating mirror, which was focused on the line S–S (see Figure 1.7) in order to observe the cessation of total reflection of light from the boundary of the glass block contacting the end face of the explosive charge. The resolution of the camera was 20 mm/μs and the width of the slit was 0.051 mm. All experiments were carried out at an ambient temperature in the range of 15–20 °C.

Figure 1.8 shows the wave configuration in the experimental setup, which was reconstructed based on the results of the measurements. An elastic wave in beryllium was usually observed as an initial partial change in the reflection of light in the glass block. A plastic wave, which arrived later, led to complete disappearance of reflection. Thus, it was possible to

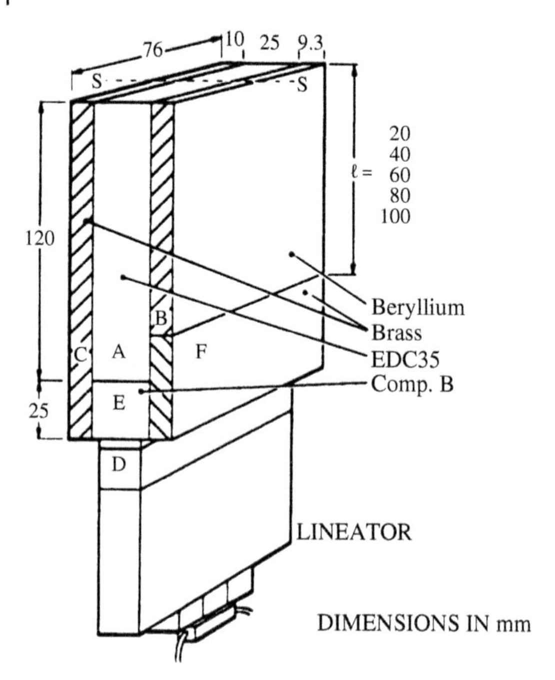

Figure 1.7 Experimental assembly.

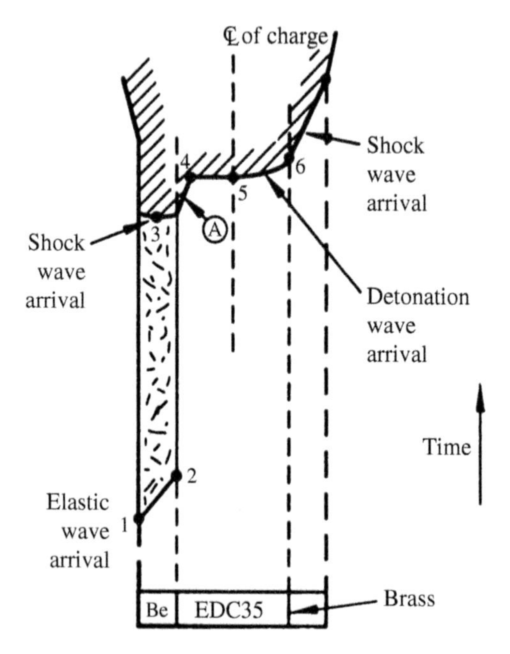

Figure 1.8 Wave configuration in the experimental assembly. A, advanced wave in EDC35.

observe that the shape of the detonation wave front was clearly affected by the wall material and, consequently, the front was asymmetric near the central line. Arriving the last was a shock wave, propagating along the surface of the bronze wall.

Based on the experimental results, the authors made the following conclusions. The detonation velocity in the inert interface with EDC35 depends on the choice of material. In the case of bronze, the detonation velocity was reduced by $0.6 \pm 0.1\%$, while in the case of beryllium it increased by $1.2 \pm 0.1\%$. An increase of the detonation velocity in the case of a beryllium wall is associated with a thin layer of EDC35, which was subjected to a preliminary impact loading by an outrunning wave in the wall before it was reached by the detonation front. The estimated amplitude of the shock wave in beryllium was 2.8 ± 0.6 GPa. In this case, it refers to the amplitude of a plastic wave, which propagates at a speed of 8.22 km/s. The estimated amplitude of the shock wave in EDC35 was 1.5 ± 0.5 GPa. The low amplitude of the wave in beryllium led to formation of a two-wave structure – the elastic wave propagated at the speed of 12.9 km/s and had an amplitude of about 0.38 GPa. Although this wave could be measured at a distance of at least 30 mm before the shock front, it did not affect the behavior of the EDC35. This fact is probably related to the low value of the Poisson's ratio for beryllium.

Authors of [21] described the results of experimental and numerical studies of the interaction of oblique detonation waves at the explosive/solid material interface for several incidence angles. The authors described an experimental method, which allowed to analyze the two-dimensional flow pattern in the case of oblique interaction. In the first part of the study, the interaction at the TATB/copper interface was studied. The experimental results showed that the solutions have a continuous nature. Outside the zone of regular interaction, there is a discrepancy between the results obtained using the hydrodynamic theory and the numerical simulation results using the two-dimensional Lagrange method.

Let us dwell in more detail on the second part of the study [21], in which the interaction of detonation waves with an elastoplastic alumina-based material was investigated. Two powerful heterogeneous explosives were used (TATB-based – composition T2 and octogen [HMX]-based – composition X1). In this study, the velocity of sound in the wall material exceeded the detonation velocity of the explosive charge.

The critical angles η_{cr} at which the flow velocity was equal to the sound velocity in the wall material were calculated using the shock polar method. For the Al_2O_3/T2 pair $\eta_{cr} = 46°$, while for the Al_2O_3/X1 pair $\eta_{cr} = 55°$.

The scheme of experimental assembly is shown in Figure 1.9. Prismatic explosive charges initiated by a plane wave generator via an intermediate

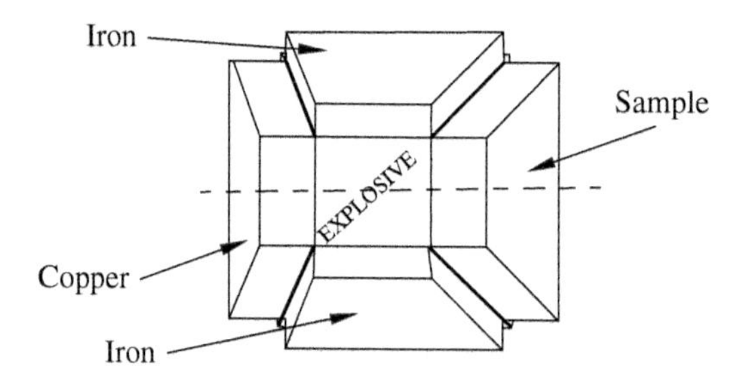

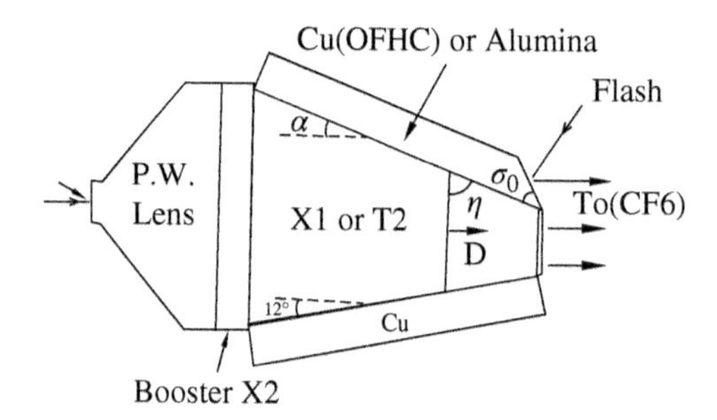

Figure 1.9 The scheme of the experimental assembly.

detonator were used. Such setup ensured formation of stable flat two-dimensional flows. One of the sides of experimental samples was machined and polished at an angle $\sigma_0 = \eta$.

An optical method with a streak camera was used to visualize the experiment, which allowed photographing the glow at the time when detonation front reached the free surface of the explosive charge simultaneously with the output of the shock wave on the polished surface of the analyzed sample. This allowed measuring the angle at which the shock wave propagated relative to the interface in the sample. Several experiments were carried out with angles η both smaller and larger than critical. The angle η varied in the range from 40° to 90°. While η was less than η_{cr}, an elastic precursor and a plastic shock wave connected with a detonation wave in the explosive charge stood out. For the cases when $\eta > \eta_{cr}$, the elastic precursor was ahead of the detonation wave, while the plastic wave remained attached to it. The angle between the elastic precursor and the interface approached 90° regardless of the initial angle of incidence. However, there was a change in the shape of the detonation wave front near the interface. This phenomenon (outrunning or slowing down) depended on the nature of the explosive and on the initial angle η. For $\eta = 76°$, the curvature of the front appeared independently of the type of explosive. However, if η approached 90°, the effect of the alumina wall was different for compositions X1 and T2. While for composition X1, a local acceleration of the detonation wave near the interface was observed, a deceleration was observed for T2, in a similar manner as in the case of the copper wall. Additional experiments carried out with each composition of explosives for charges having other weights and lengths have shown the presence of similar wave configurations. This can serve as an evidence of the quasistationary nature of the flow near the localization of the detonation wave on the interface. When the detonation wave passes approximately 100 mm, the local acceleration (or deceleration) does not change anymore.

In [22, 23], it was experimentally and theoretically shown that if the speed of sound in an inert wall exceeds the detonation velocity of the explosive charge, then an oblique compression wave without a shock front propagates in the wall.

One of the earliest observations of the processes of detonation interaction with shells was described in [24], where Karpukhin et al. studied the phenomenon of initiation of a chemical reaction front ahead of a detonation wave under the action of an elastic wave, which outruns the detonation process in cylindrical shells made of glass or quartz. Fine ammonium perchlorate powder (with particle size in range of 1–5 μm) with various explosive and nonexplosive additives (TNT, aluminum powder, polymethyl methacrylate [PMMA], etc.) was used as the object of study. As a rule, charges of gravimetric density (relative density $\delta = 0.6$) were investigated. Most of the experiments were carried out using glass shells with an internal diameter d in the range from 5 to 20 mm and with wall thickness from 0.5 to 5.0 mm. The length of charges l varied from 40 to 250 mm. The initiation was performed by an intermediate RDX charge of gravimetric density or directly using the detonator via PMMA barrier with thickness of 4–6 mm. The process was recorded from the lateral surface by using a GFR-2 high-speed camera.

The normal detonation velocities of explosives (2.5–3.2 km/s) always turned out to be much lower than the velocity of the elastic wave propagation in the shell. As a result, the elastic wave was significantly ahead of the detonation propagation front in the charge and could directly affect the layers of explosives adjacent to the shell. When the shock wave left the barrier, two brightly illuminated reaction fronts were simultaneously observed in the experimental charge with different velocities, D_1 and D_2, where D_2 is the velocity of normal detonation, and $D_1 > D_2$. The velocity of the first emission front D_1 could reach 4.0–4.5 km/s. The two glowing fronts diverged at an angle from one point and formed a triangle at some length of the photographic registration, inside which one could observe clearly visible transverse luminous tracks. If the two glowing fronts propagated to the end of the charge, then when observing the process from the end, one could see an outrunning glowing ring in the peripheral region. The lead of the first front over the second one could reach 5–7 μs or 15–22 mm. When the first glowing front passed 80–100 mm of the length of the charge, the emission pattern began to change. The velocity of the first front D_1 began to gradually approach the velocity of the detonation front D_2, its glow became blurred, and the gap between the fronts remained constant (about 5–7 μs), and it was oversaturated with transversal tracks.

According to the authors, as the elastic wave passes along the shell, regardless of whether it is caused by the detonation of the initiating charge or the explosive transformation of the substance itself, a lateral compression wave propagates into the cylindrical charge, which at sufficient intensity can serve as a source of initiation of chemical reactions in the layers adjacent to the inner surface of the shell. The effect of the lateral compression wave on the material placed inside the shell leads to an adiabatic compression of the gas inclusions, which results in local heating that, after a certain period of induction, can cause ignition of the surrounding particles of the explosive mixture. These are the reaction centers that create transversal tracks on the photogram inside the triangle.

The effect of shells fabricated from materials with sound velocity exceeding the detonation velocity has been often investigated for mixed explosives with a low detonation velocity [25]. This allows using shells fabricated from ordinary metals, for example, aluminum. The results observed in such cases are typical for processes in low-sensitivity explosives.

1.4 Summary

The examples shown demonstrate nonstationary effects in detonation processes in the presence of wave perturbations that outrun the detonation front. In this regard, it is of interest to study similar processes with explosives having sufficiently high velocity of stationary detonation. In this case, one obviously needs to have an appropriate choice of inert materials used in the experiments. Such materials typically are ceramics, the speed of sound in which significantly exceeds the detonation velocity of explosives most frequently used for scientific and practical purposes. The choice of appropriate inert materials is justified in the next chapter.

References

1 Urbański, T. (1926). Fotograficzne badania przebiegu detonacji kruszących materiałów wybuchowych. *Roczniki Chemii* 6: 838–847. [in Polish].
2 Urbański, T. (1977). Some phenomena of detonation of solid explosives. *Archiwum termodynamiki i spalania* 8 (1): 5–12.
3 Dubnov, L.V. and Khotina, L.D. (1966). Channel effect mechanism in the detonation of condensed explosives. *Combustion, Explosion and Shock Waves* 2 (4): 59–63.
4 Dubnov, L.V., Bakharevich, N.S., and Romanov, A.I. (1988). *Industrial Explosives.* Moscow: Nedra [in Russian].
5 Goldbinder, A.I. and Tyshkevich, V.F. (1964). On the channeling effect in the detonation of explosives. *Dokladi Akademii Nauk SSSR* 156 (4): 905–908. [in Russian].
6 Golbinder, A.I. and Tyshkevich, V.F. (1967). Further investigation of the channel effect. In: *Theory of Explosives: A Collection of Articles* (ed. K.K. Andreev), 349–368. Moscow: Vysshaya Shkola [in Russian].
7 Dremin, S.D., Savrov, V.S., Trofimov, K.K. et al. (1970). *Detonation Waves in Condensed Matter.* Moscow: Nauka [in Russian].
8 Sumiya, F., Tokita, K., Nakano, M. et al. (1999). Experimental study on the channel effect in emulsion explosives. *Journal of Materials Processing Technology* 85: 25–29.
9 Woodhead, D.W. (1947). Velocity of detonation of a tubular charge of explosive. *Nature* 160 (4071): 644.
10 Woodhead, D.W. (1959). Advance detonation in a tubular charge of explosive. *Nature* 183 (4677): 1756–1757.
11 Woodhead, D.W. and Titman, H. (1965). Detonation phenomena in a tubular charge of explosive. *Explosivstoffe* 13 (5): 113–123. (6): 141–155.
12 Ahrens, H. (1965, 1967). Über den detonationsvorgang bei zylindrischen sprengstoff-ladungen mit axialer höhlung. *Explosivstoffe* 13: 124–134. 155–164; 180–198; 267–276; 295–309; 15: 121–129; 145; 175–185.
13 Zagumennov, A.S., Titova, N.S., Fadeenko, Y.I., and Chistyakov, V.P. (1969). Detonation of elongated charges with cavities. *Journal of Applied Mechanics and Technical Physics* 10 (2): 246–250.
14 Johansson, C.H. and Persson, P.A. (1970). *Detonics of High Explosives.* London/New York: Academic Press.
15 Mitrofanov, V.V. (1975). Ultra-high-speed detonation in charges with longitudinal channels. *Combustion, Explosion and Shock Waves* 11 (1): 63–70.
16 Brailovsky, I. and Sivashinsky, G. (2014). Precursors in two-phase detonation: an elementary model for the channel effect. *Combustion Theory and Modelling* 18 (1): 117–147.

17 Tanguay, V. and Higgins, A.J. (2004). The channel effect: coupling of detonation and the precursor shock wave by precompression of the explosive. *Journal of Applied Physics* 96 (1): 4894–4902.
18 Merzhievskii, L.A., Fadeenko, Y.I., Filimonov, V.A., and Chistyakov, V.P. (1976). Acceleration detonation propagation in charges with a litium-filled hollow. *Combustion, Explosion and Shock Waves* 12 (2): 205–211.
19 Merzievski, L.A., Fadeenko, Y.I., and Chistjakov, V.P. (1977). Detonation of cylindrical charge with lithium-filled cavity. *Acta Astronautica* 4: 459–496.
20 Eden, G. and Belcher, R.A. (1989). The effects of inert walls on the velocity of detonation in EDC 35, an insensitive high explosive. *Proceedings of the 9th International Symposium on Detonation*, Portland, USA (28 August–1 September 1989). Arlington: Office of Naval Research.
21 Aveille, J., Carion, N., Vacellier, J. et al. (1989). Experimental and numerical study of oblique interaction of detonation waves with explosive/solid material. *Proceedings of the 9th International Symposium on Detonation*, Portland, USA (28 August–1 September 1989). Arlington: Office of Naval Research.
22 Neal, T.R. (1976). Perpendicular explosive drive and oblique shocks. *Proceedings of the 6th Symposium on Detonation*, Coronado, USA (24–27 August 1976). Arlington: Office of Naval Research.
23 Sharpe, G.J. and Bdzil, J.B. (2006). Interactions of inert confiners with explosives. *Journal of Engineering Mathematics* 54: 273–298.
24 Karpukhin, I.A., Balinets, Yu. M., Bobolev, V.K. et al. (1977). Initiation of fast chemical reactions in solid mixed explosives by elastic wave in cylindrical shell. *Proceedings of the Soviet Union Meeting on Detonation*, Chernogolovka, USSR (22–26 August 1977). Chernogolovka: IPCP RAS. [in Russian].
25 Short, M. and Jackson, S.I. (2015). Dynamics of high sound-speed metal confiners driven by non-ideal high-explosive detonation. *Combustion and Flame* 162: 1857–1867.

2

Phenomena in High Explosive Charges Containing Rod-Shaped Inert Elements

Describing shock wave processes, one typically uses relations derived under the assumption that the wave is plane and propagates in a space that is unbounded in the direction transverse to the front motion. This means that a uniaxial deformation is realized under the shock compression. In practice, specimens and elements of various designs always have finite dimensions, and shock compression can be accompanied by transverse deformations. This is particularly noticeable in the case of shock waves propagation in rods, the transverse dimensions of which are much smaller in comparison to the longitudinal ones. In this case, the initially specified impulse can be considerably transformed, which in the case of interaction of detonation waves with rods of high-modulus materials will lead to a number of effects, some of which are described in this chapter.

2.1 "Smoothing" of Shock Waves in Silicon Carbide Rods

2.1.1 Experiments with Ceramic Rods

Depending on the amplitude of the load achieved during the shock loading of solids, waves having a single-wave structure (elastic wave, elastic–plastic compression) or a two-wave structure (elastic precursor and a plastic wave) can propagate in them [1]. In the case of high-modulus materials, the Hugoniot elastic limit for which is sufficiently high, the amplitudes of these elastic waves can reach tens of gigapascals (GPa). For example, the Hugoniot elastic limit of silicate glass is in the range of 6–10 GPa [2], while in silicon carbide it can reach 16 GPa [3]. Theoretical analysis and experimental data show that the evolution of shock waves occurs even for elastic waves propagating in rods. This is caused by the

Explosion Systems with Inert High-Modulus Components: Increasing the Efficiency of Blast Technologies and Their Applications, First Edition. Igor A. Balagansky, Anatoliy A. Bataev, and Ivan A. Bataev.

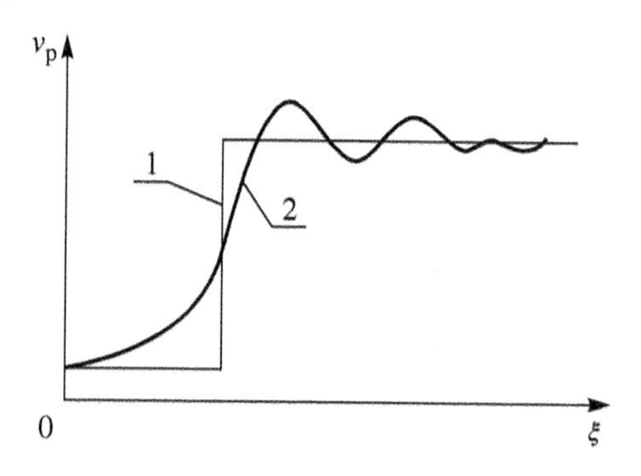

Figure 2.1 Evolution of a shock front: 1, initial impulse and 2, "smoothing" impulse.

influence of cross-section inertia due to the finiteness of its transverse dimensions [4, 5]. For example, in the case of a stepped load in an elastic rod, as the time flows, the original impulse is becoming blurred [6]. This effect is schematically illustrated in Figure 2.1 [6], where the change in the front shape is shown in the "velocity–dimensionless time" coordinates.

The size of the "blurred" (smoothened) region d in the first approximation can be estimated using equation [6]:

$$d \approx \sqrt[3]{r^2 x}.$$

Here r is the radius of the rod; x is the distance traveled by the wave front in the rod. In this case, the propagation velocity of the region with the highest amplitude corresponds to the rod sound velocity in the material (see Appendix A). An estimate of the front width of a stationary elastoplastic wave in the case of a single-wave configuration can be obtained based on a description of the relaxation properties of the material [7].

Some examples of the experimentally observed effect of the shock wave blurring may be found in [8–10], where it was found in shock-loaded square bars of silicon carbide. In these experiments, the shock wave profiles generated by explosion were measured using manganin gauges (see Appendix B). A scheme of the experimental assembly is shown in Figure 2.2.

The rods of square cross section (20 × 20 mm) and lengths of 20.0, 40.0, and 77.5 mm made of self-bonded silicon carbide were used in the experiments. The sample density was 3.08 g/cm^3, the porosity was lesser than 2%, the free silicon content was 9.5%, and the free carbon content was 3%. Before the experiments, flat ends of all rods were polished. The estimated value of the Hugoniot elastic limit was 10 GPa. Shock waves in the rods were induced by the detonation of cylinder-shaped explosive

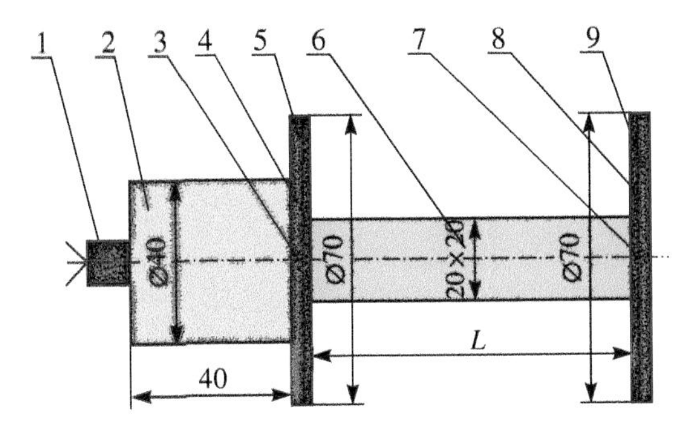

Figure 2.2 The scheme of the experimental assembly: 1, detonator; 2, hexogen (RDX) charge ($\rho_0 = 1.60\,g/cm^3$); 3, 7, manganin gauges; 6, ceramic rod; and 4, 5, 8, 9, copper screens.

charges with a diameter and height of 40 mm, which were prepared by pressing phlegmatized hexogen (RDX) to a density of $1.60\,g/cm^3$. Between the explosive charge and the silicon carbide rod, thin copper screens were placed. The first manganin gauge was mounted between these screens. Copper screens and the second manganin gauge were also placed at the opposite end of the rod. The gauges had a thickness of 20 μm, and they were isolated from copper screens by fluoroplastic spacers with a thickness of 80 μm. The gauges were used to measure pressure profiles of the waves propagating along ceramic rods of different lengths. The accuracy of time measurements was not worse than 0.01 μs, the accuracy of pressure measurements was not worse than 5%, and time resolution in the experiments was about 0.1 μs.

The main results of measurements with manganin gauges in copper screens are summarized in Table 2.1. The results, recalculated for the values of the parameters in ceramics and explosives, are given in Table 2.2. Figures 2.3–2.5 show the wave profiles measured by manganin gauges. The solid lines are the profiles recorded by the first and the second gauges, and the dashed lines correspond to profiles recalculated for parameters in ceramics.

Figure 2.6 shows the wave profiles calculated from experiments in a ceramic rod at various distances from the loaded end.

The propagation of a shock wave along the ceramic rod was also investigated by high-speed optical recording using the ZHLV-2 camera in the frame mode (Figure 2.7). The time step between frames was 2.66 μs. The sequence of the frames presented in Figure 2.7 illustrates the flow of ceramic particles moving at high velocities. The experimental data indicates that the propagation velocity of the compression impulse along the length of the ceramic rod remains constant. This velocity does not depend on

Table 2.1 Main results of measurements with manganin gauges.

Experiment number	Length of rod l, mm	Thickness of copper screens				Maximum pressure in the first gauge P_1, GPa	Maximum pressure in the second gauge P_2, GPa	Time between the first and the second gauges Δt, μs
		h_1, mm	h_2, mm	h_3, mm	h_4, mm			
1	20.0	1.4	1.4	1.4	5.0	36.2	8.2	2.44
2	40.0	1.4	1.4	1.4	5.0	35.6	5.7	4.32
3	77.5	3.0	2.0	2.0	3.0	32.0	2.5	7.92

Table 2.2 Results for ceramics and high explosives.

Length of rod l, mm	Wave height in the rod P, GPa	Wave height in the passive HE charge P_{HE}, GPa	Experimental value of front fuzziness area d, mm	Wave velocity in the rod D, km/s	Calculated value of front fuzziness area d_c, mm
10.0	~20.0	~8.0	—	—	10.8
20.0	7.5	2.2	3.3	11.0	13.7
40.0	5.7	1.6	5.3	10.9	17.2
77.5	2.5	0.7	11.0	11.1	21.5

the compression amplitude and is equal to approximately 11 km/s, which corresponds to the elastic wave speed in the rod of this material.

One may observe a strong smoothing of the wave front as it propagates along the rod that corresponds to the theoretical concept regarding the impossibility of existence of an elastic shock wave in a body of finite transverse dimensions.

Numerical estimates of the size of the front blur (d) are consistent with the measured experimental values. For instance, for a 77.5 mm rod, the experimentally measured value of d was 11.0 mm, the calculated value of d was 21.5 mm.

It should be noted that the amplitude of the wave is significantly attenuated at the initial 20 mm of the sample length (more than four times) and only a relatively small attenuation occurs during its subsequent propagation (at the length of 57.5 mm, it is getting only three times smaller), which agrees with results of [11]. Apparently, this happens due to the decrease in the wave amplitude below the Hugoniot elastic limit at the initial part of the rod and its subsequent propagation in the elastic region.

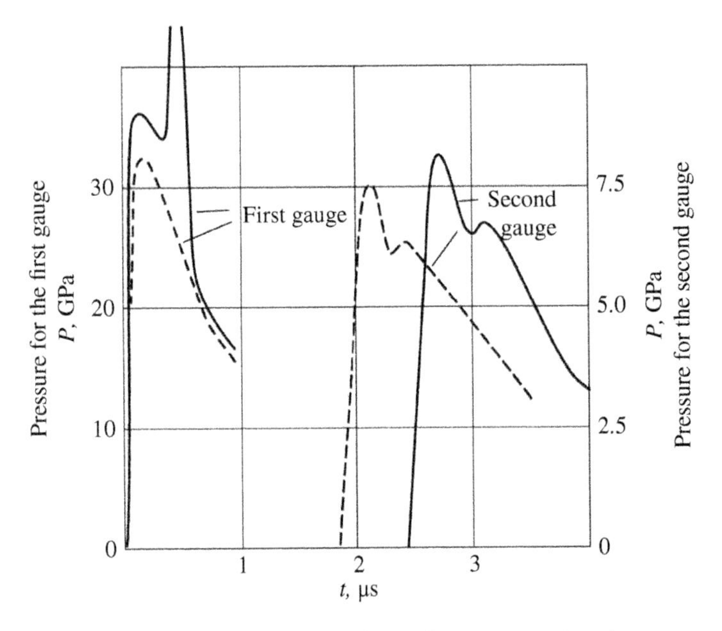

Figure 2.3 Wave profiles at the end faces of 20 mm ceramic rod.

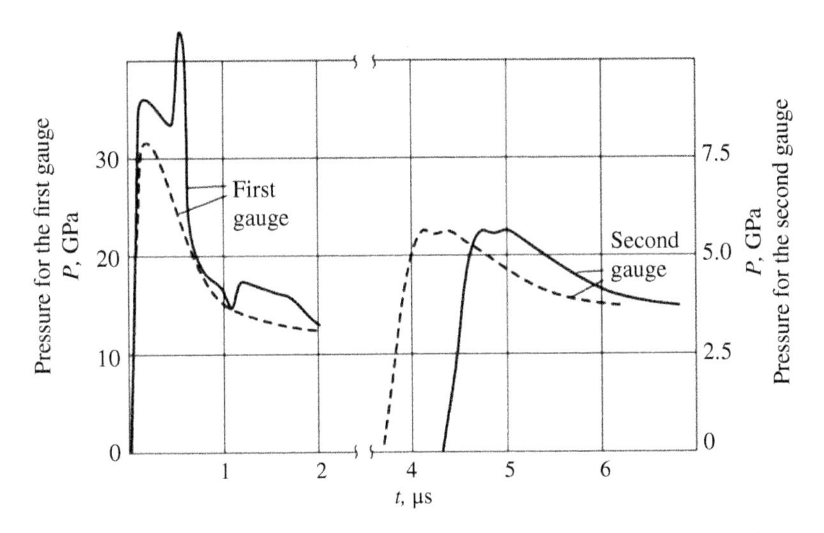

Figure 2.4 Wave profiles at the end faces of 40 mm ceramic rod.

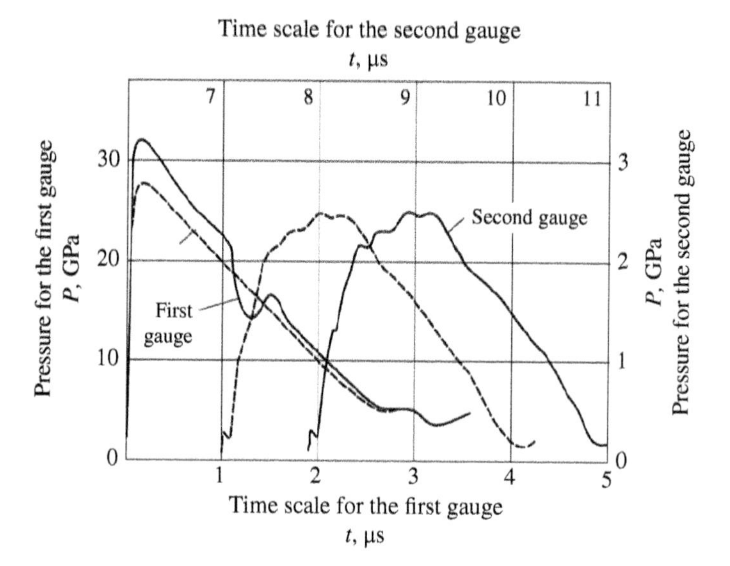

Figure 2.5 Wave profiles at the end faces of 77.5 mm ceramic rod.

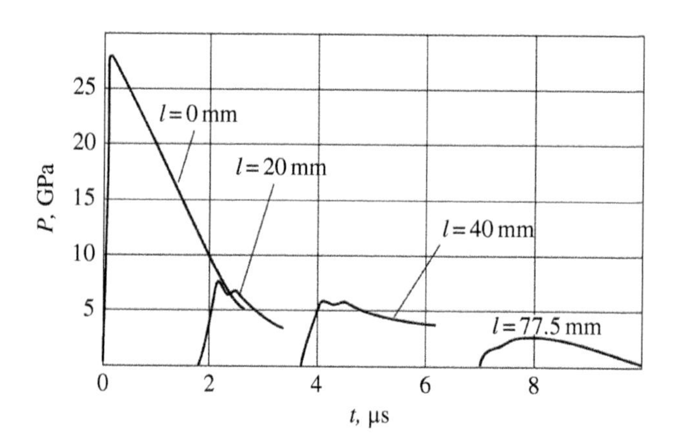

Figure 2.6 Wave profiles in a ceramic rod at various distances from the loaded end.

2.1.2 Numerical Simulation of Shock Wave Propagation in Silicon Carbide Rods

The experimental information presented in Section 2.1.1 was compared to the results of a numerical simulation of the shock wave evolution process in a silicon carbide rod, carried out using ANSYS AUTODYN [12]. The corresponding problem was solved in a 2D-axis symmetry formulation. The configuration of the computational area corresponded to the

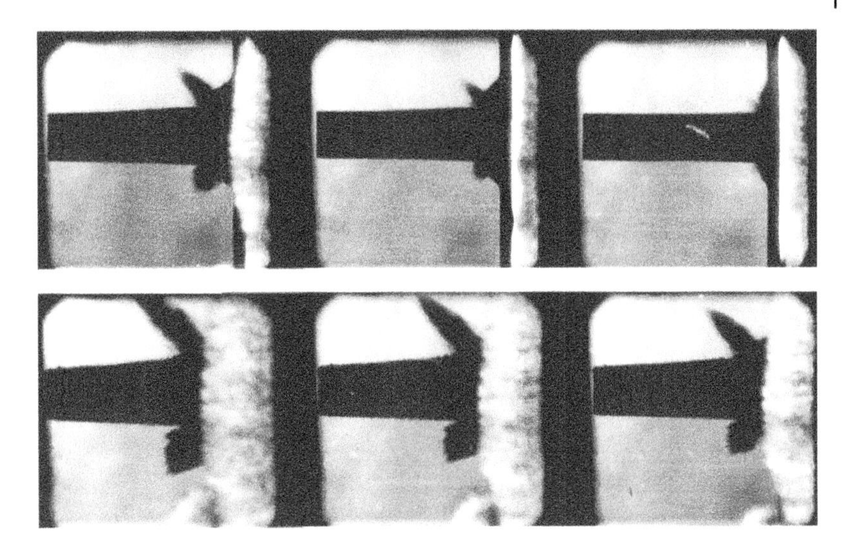

Figure 2.7 The process of shock wave loading of a ceramic rod by explosion of a phlegmatized RDX charge with a height and diameter of 40 mm.

experiments, but instead of rods of square cross section, round bars with a diameter of 22 mm having the same cross-sectional area were used in the simulation. The numerical solution of the corresponding mathematical problem was performed on a rectangular Eulerian mesh with a spatial resolution of 4 cells/mm. On the left, right, and upper boundaries of the mesh, the flow out (Euler) boundary conditions were specified, which ensured the free flow of matter and the passage of waves beyond the calculated region without reflections from the boundaries.

The parameters of the governing relationships for materials were selected from the standard AUTODYN equation of state (EOS) library, except for the parameters for silicon carbide. The governing equations and their parameters for silicon carbide were selected from a series of preliminary test calculations, which were compared to the experimental data. The best results were obtained when the material was described using the model of an ideal elastoplastic behavior without considering possible failure.

The spherical part of the stress tensor was calculated using the Rankine–Hugoniot relations, and the deviatoric component was calculated in accordance with the Hooke's law and the law of plastic flow associated with the Mises criterion according to the Wilkins procedure [13].

Table 2.3 summarizes the governing relations used for the calculation in the terminology of the AUTODYN software. Table 2.4 presents the constants describing the behavior of silicon carbide, which were selected according to the results of the test calculations. The main results of calculations are shown in Table 2.5.

Table 2.3 Relations used for the calculation of all involved substances.

Material name	Equation of state	Strength	Failure
AIR	Ideal gas	None	None
COPPER1	Shock	Johnson Cook	None
OCTOL	JWL	None	None
SiC-IPM	Shock	von Mises	None

Table 2.4 Constants describing the behavior of silicon carbide.

Reference density	**3.10 g/cm^3**
Gruneisen coefficient	1.25
Parameter C_1	8.00 mm/us
Parameter S_1	0.95
Strength	**von Mises**
Shear modulus	1.70 Mbar
Yield stress	0.065 Mbar
Failure	**None**

Table 2.5 Maximum pressures and times recorded by gauges for rods of different lengths.

Length of rod l, mm	Maximum pressure in the first gauge P_1, GPa	Time for the first gauge t_1, μs	Maximum pressure in the second gauge P_2, GPa	Time for the second gauge t_1, μs
20	36.6	5.30	6.48	8.20
40	36.7	5.30	4.88	9.90
80	37.1	5.30	4.18	13.40

Figure 2.8 shows the initial configuration of the computational area for a ceramic rod with a length of 20 mm. Figure 2.9 shows the wave profiles calculated for this case, recorded by the first and the second gauges.

Figure 2.10 shows the initial configuration of the computational area for a 40 mm ceramic rod. The corresponding calculated wave profiles are presented in Figure 2.11. Similar data for an 80 mm rod is shown in Figures 2.12 and 2.13.

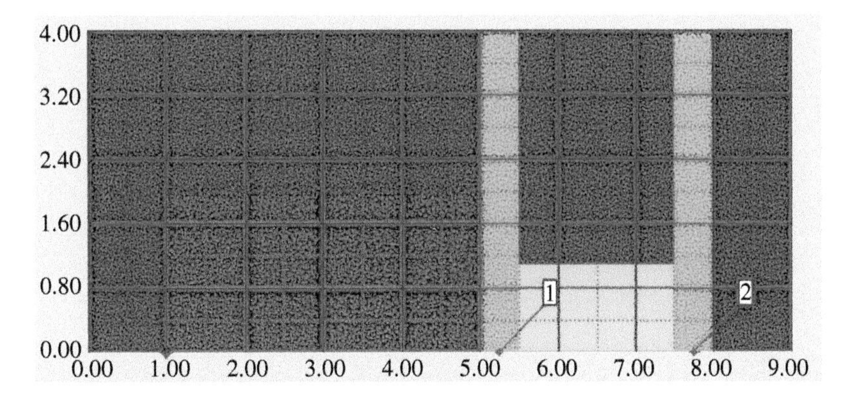

Figure 2.8 Initial configuration of the calculation area, $l = 20$ mm: 1 and 2, gauges for recording pressure dependencies on time.

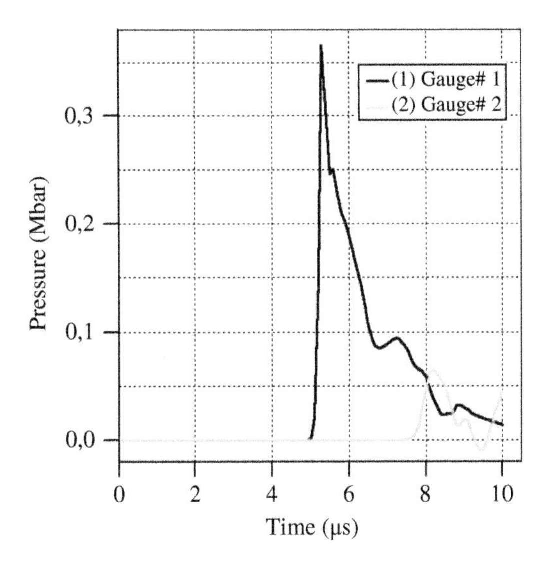

Figure 2.9 Wave profiles, $l = 20$ mm.

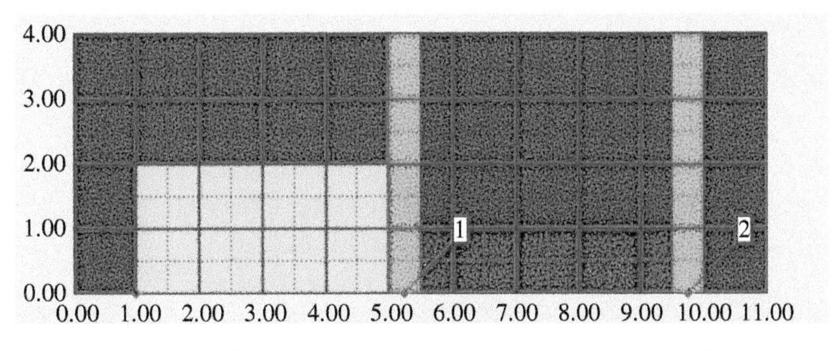

Figure 2.10 Initial configuration of the calculation area, $l = 40$ mm: 1 and 2, gauges for recording pressure dependencies on time.

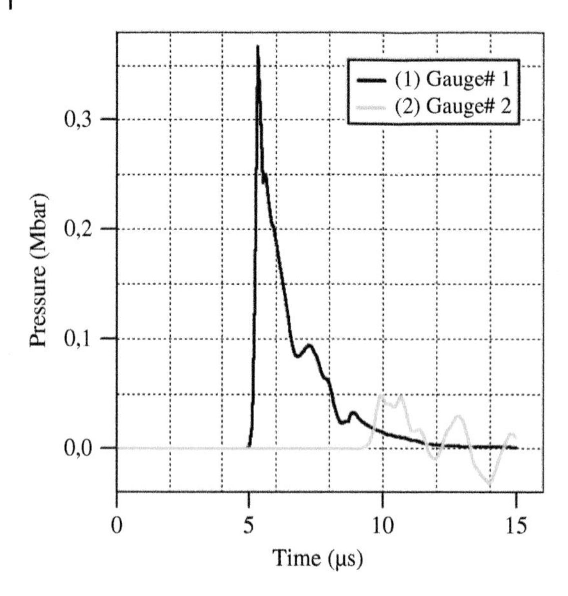

Figure 2.11 Wave profiles, $l = 40\,\text{mm}$.

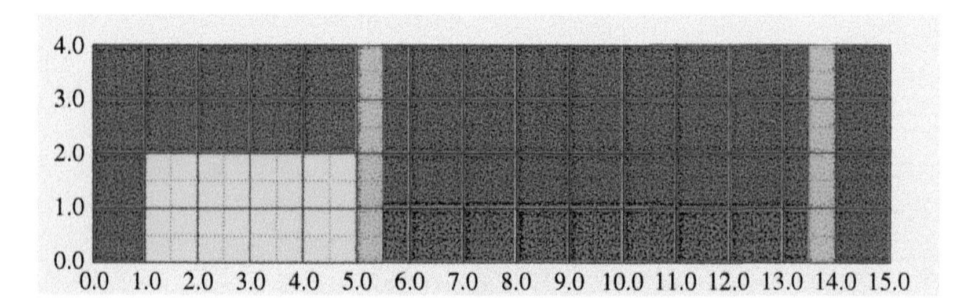

Figure 2.12 Initial configuration of the calculation area, $l = 80\,\text{mm}$.

Comparison of the simulation results to the experimental data shows that the evolution of shock impulses observed in experiments is adequately described using the selected models. The simulation helps analyze the details of these processes.

2.2 Desensitization of Heterogeneous High Explosives After Loading by Advanced Waves Passing Through Silicon Carbide Elements

According to modern concepts, the main mechanism for the development and propagation of detonation processes in heterogeneous high explosives (HE) is related to the formation and evolution of hot spots

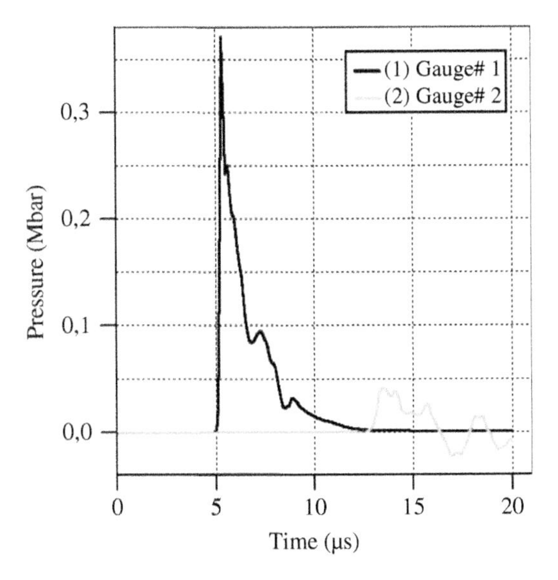

Figure 2.13 Wave profiles, $l = 80$ mm.

caused by a number of microstructural and deformation properties of explosives. These properties are very sensitive to external influences, which in this way can affect the sensitivity of explosives. The first indirect manifestations of shock desensitization were obtained in experiments with preliminary action of weak shock waves on explosive charges.

Currently, the phenomenon of desensitization of heterogeneous explosives loaded by a shock wave with a precursor of a smaller amplitude is well known [14–17]. The term "shockwave desensitization" is associated with various phenomena in explosives, such as "dead pressing," attenuation of detonation, or decrease in the reaction of explosives under multi-wave loading. Thus, the inert behavior of an explosive can be achieved at much higher pressures in the main shock wave. If the initial impulse has a precursor with an amplitude that is insufficient to initiate detonation, the pores in the heterogeneous explosive are closed and the charges do not explode even if the pressure in the main shock wave is much higher than the level of reliable initiation.

Most researchers believe that desensitization is the result of deactivation of hot spots [14–16]. There are various theories about the exact nature and role of hot spots in the initiation of explosives. There is a general agreement that under shock loading, hot spots are formed from regions of reduced density that are heated to a temperature much higher than the average temperature of the entire volume [17]. These regions react with the release of energy and detonation products.

The mysterious fact is that the hot spots ignited by a relatively weak shock do not continue to react and do not expand into the rest of the explosive. The explosion does not extend beyond a thin layer. A strong second shock sent after the first one to the preliminary compressed area does not lead to the initiation of detonation. The initiation process starts only when the second shock wave overtakes the first one and propagates via the uncompressed explosive.

For instance, in [15], it was found that the polymer-bonded explosive (PBX-9404) compressed by a shock wave to a pressure of 3.9 GPa does not detonate when loaded with the second shock wave with an amplitude of 10 GPa. Another example may be found in [18], where it was shown that in cast trinitrotoluene (TNT) preliminary compressed by a shock wave with an amplitude of about 4 GPa, the second shock wave with an amplitude of 20 GPa propagated with a substantially lower velocity than the velocity of stationary detonation. The desensitization of phlegmatized RDX and octogen (HMX) by successive shock waves was studied in detail in [19]. Similar effects were observed during the initiation of heterogeneous explosives by an isentropic wave with a front blurred in time [20, 21].

To explain the phenomenon in better detail, we show different types of initiating impulses in Figure 2.14. If the initial impulse has a precursor whose amplitude is insufficient to initiate a detonation, or if the high compressive pressure is reached insufficiently fast, then when the pores are closed in a heterogeneous explosive, the necessary conditions for detonation initiation are not achieved and the charge does not detonate even when the final compressive pressure exceeds the level of stationary detonation pressure.

Thus, it seems natural to investigate the process of detonation initiation by smooth shock waves formed in ceramic rods.

2.2.1 The Experiments on Detonation Transmission

To study the effect of shock front transformation upon propagation in ceramic rods on the initiation of detonation, we used the experimental setups shown in Figures 2.15 and 2.16.

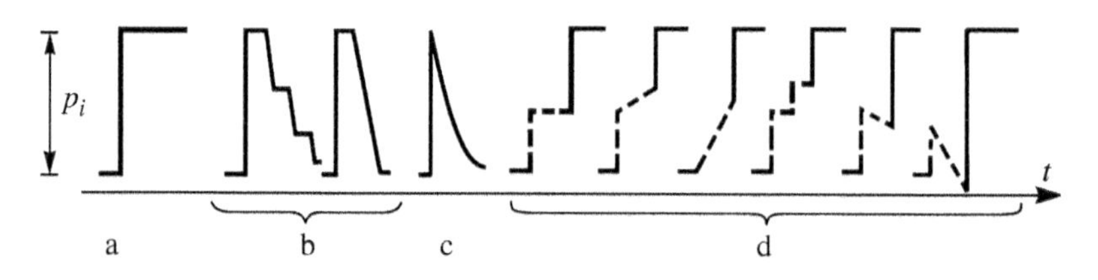

Figure 2.14 Initial impulses of various types: a, stepwise; b, short; c, triangular; and d, "with a precursor."

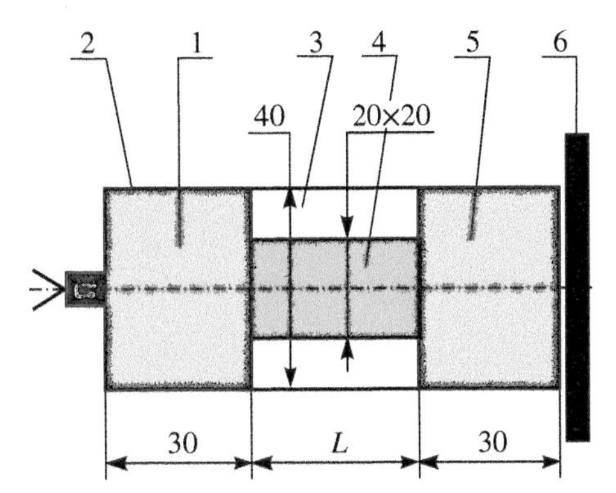

Figure 2.15 Scheme of experimental assembly: 1, active charge; 2, shell; 3, inert matter; 4, silicon carbide rod; 5, passive charge; and 6, polymethyl methacrylate (PMMA).

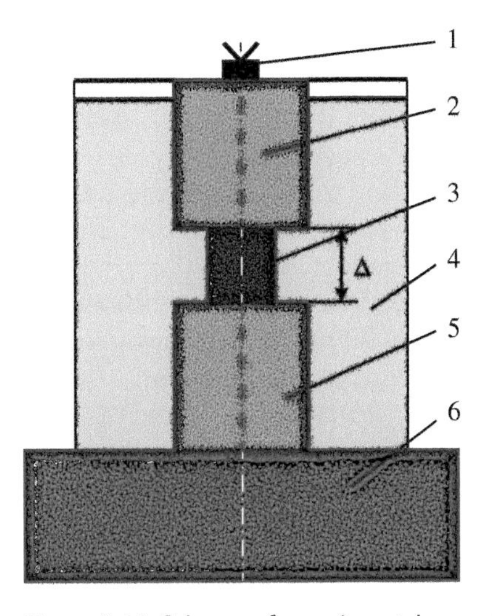

Figure 2.16 Scheme of experimental assembly: 1, electric detonator; 2, active high-explosive (HE) charge; 3, inert insert (shaped as a 20 mm × 20 mm square prism); 4, degassed water; 5, passive HE charge; and 6, identification steel specimen. (*Source:* From Balagansky and Stepanov [22]. Reprinted with permission of Springer Nature.)

In the first series of experiments, the possibility of initiating an HE charge of phlegmatized RDX by a shock wave passing through self-bonded silicon carbide rods was investigated [23]. The silicon carbide rods with 20 × 20 mm cross section and various lengths were placed between the active and passive HE charges. Both the active and the passive charges were produced from phlegmatized RDX cylinders with a

Table 2.6 Conditions and results of experiments.

Experiment number	Inert matter	Length of ceramic rod, mm	Detonation
1	Epoxy resin	80	No
2	Air	40	No
3	Air	20	No
4	Air	10	No

diameter of 40 mm, a height of 30 mm, and a density of 1.64 g/cm^3. The scheme of the experimental assembly is shown in Figure 2.15.

The experiments were recorded by the SFR-2M rotating mirror camera in the streak mode by scanning of the glow of the passive charge end face. The experimental conditions and results are given in Table 2.6.

The results of these experiments indicate that the wave smoothened due to the propagation along the ceramic rod has completely lost its initiating ability. The detonation of the passive charge was not observed even when the length of a rod was only 10 mm. It should be noticed that there was also no subsequent initiation of the passive charge by the arriving air shock wave and detonation products of the active charge, i.e. the desensitization of the passive charge was observed. In comparison, according to the data published in [24], under similar conditions the distance for reliable transmission of detonation through air for this type of explosive is about 75 mm, while the distance for detonation transmission through water, steel, and aluminum is approximately 15–20 mm.

In the next series of experiments (Figure 2.16), we investigated the detonation transmission through water after loading the passive charge of explosive with an outrunning wave that passed along the rods of different length manufactured from different materials [22, 25–30]. In these experiments, fluoroplastic, copper, and self-bonded silicon carbide were used as materials of inert inserts. The inserts had square cross section (20 × 20 mm) and various lengths. They were located between the active and passive charges. Inert inserts were mounted using vacuum lubricant "Ramzai" to prevent any clearance between HE charges and inserts. The following considerations were taken into account to select the materials of the inserts. Inert inserts made of fluoroplastic should not lead to desensitization of the passive charge. Besides, their use makes it possible to account the effect of the decrease in the effective cross sections of the active and passive charges on the process of detonation transfer. Copper was chosen since its dynamic rigidity practically coincides with the dynamic rigidity of silicon carbide. The porosity of silicon carbide used in the experiments was less than 2%, the rod sound velocity was 11.0 km/s.

The cast TG-40 (Russian analog of Composition B) with a density of 1.65 g/cm^3 was used for both active and passive charges. High-voltage detonators ATED-15 were used to initiate the active charge. To determine the presence or absence of detonation, we placed a massive steel specimen at the rear end of the passive charge. The assembly was fastened with three strips of laminated plastic with a thickness of 1 mm and a width of 10 mm along the entire length of the charges and fixed with Scotch tape (not shown in the figure). We ensured a complete absence of air bubbles in the water. In most experiments, the front end of the active charge was 5 mm above the water level, except for experiments with optical registration. In the latter cases, the charge was submerged in water up to its upper surface. The choice of water as a transmission medium was due to its transparency at all stages of the process. In contrast to experiments conducted in air, the process of detonation transmission in this case was completely determined by wave phenomena, and the transmission of detonation due to explosion products of the active charge was excluded.

At the first stage, the maximum distances at which a reliable detonation transfer takes place and the minimum distances at which the initiation of the passive charge through water occurs (i.e. without inert insertions) were determined. At the second stage, the distances for the transmission of detonation and crash of the initiation of the passive charge through water with an inert insert of fluoroplastic were found. At the third and the fourth stages, these distances were determined for copper and self-bonded silicon carbide inserts, respectively. The experimental results are summarized in Table 2.7. Table 2.8 shows the transmission and detonation crash distances for different materials of inert inserts determined from the data presented in Table 2.7.

The results indicate that loading of the passive charge by the advanced wave via copper and ceramic rods leads to a significant desensitization of the passive charge of TG-40. All other things being equal, the detonation crash distance for copper inserts was 74%, and for silicon carbide inserts, 60% of the detonation crash distance in water with inserts of fluoroplastic. It may be noted that while in the copper inserts a leading shock wave propagated during the loading, a compression wave with a blurred front propagated in the ceramic rod. This indicates that the wave with a blurred front has a greater desensitizing ability.

According to Section 2.1.1, one may estimate the parameters of waves in the passive charge of explosives, which have preliminarily passed through the self-bonded silicon carbide rods. These parameters together with the characteristics of the waves in the rods are given in Table 2.9. It may be concluded that when a length of the ceramic rod was 20 mm, the amplitude of the wave in the passive charge was approximately 2 GPa, and the region of the front blurring was approximately 3 mm.

Table 2.7 Presence of detonation in a passive HE charge.

Experiment number	Material of inert insert	Length of inert insert or distance between HE charges Δ, mm	Detonation in a passive HE charge
1	No	40.0	Yes
2, 3	No	50.0	Yes
4, 5	No	55.0	No
6	No	60.0	No
7	Fluoroplastic	40.0	Yes
8	Fluoroplastic	45.0	Yes
9, 10	Fluoroplastic	50.0	No
11, 12	Copper	20.0	Yes
13, 14, 15	Copper	30.0	Yes
16	Copper	35.3	Yes
17	Copper	35.4	No
18	Copper	37.0	No
19, 20	Copper	40.0	No
21, 22	Silicon carbide	19.5	Yes
23	Silicon carbide	19.6	Yes
24	Silicon carbide	24.8	Yes
25	Silicon carbide	26.5	Yes
26	Silicon carbide	30.4	No
27	Silicon carbide	31.0	No

Table 2.8 Distances of reliable transmission and crash of detonation.

Material of inert insert	Distance of reliable detonation transmission, mm	Crash distance of detonation transmission, mm
Without inert insert	50	55
Fluoroplastic	45	50
Copper	30	37
Silicon carbide	26	30

Source: From Balagansky et al. [30]. Reprinted with permission of World scientific Publishing Company.

Table 2.9 Parameters of waves in rods and passive HE charges.

Length of rod l, mm	Wave height in the rod P, GPa	Wave height in the passive HE charge P_{HE}, GPa	Experimental value of front fuzziness area d, mm	Wave velocity in the rod D, km/s
10.0	~20.0	~8.0	—	—
20.0	7.5	2.2	3.3	11.0
40.0	5.7	1.6	5.3	10.9
77.5	2.5	0.7	11.0	11.1

Source: From Balagansky et al. [30]. Reprinted with permission of World scientific Publishing Company.

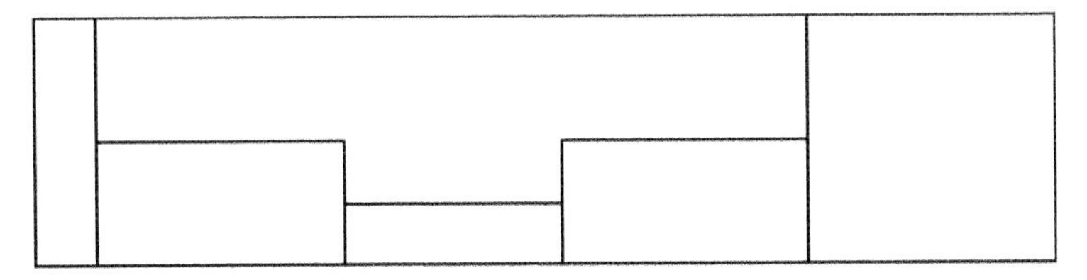

Figure 2.17 Configuration of the calculation area. (*Source:* From Balagansky and Stepanov [22]. Reprinted with permission of Springer Nature.)

2.2.2 Modeling of the Detonation Transmission Process Under Initiating Through Inert Inserts

In this section, we present the results of numerical simulation of wave processes in the experiments described in Section 2.1.2. The setups that were used in simulation were as close as possible to the corresponding experimental setups. The simulation was performed using ANSYS AUTODYN software.

The configuration of the calculation area shown in Figure 2.17 reproduced the geometry of the experimental setup (see Figure 2.16). Because of symmetry, only the upper part of the section is shown, while the lower boundary is the axis of symmetry. The spatial resolution of the regular Eulerian mesh was 10 cells/mm. The choice of the governing relationships and their parameters for inert materials is described earlier in Section 2.1.2. To simulate the behavior of the passive charge of explosive (TG-40), a Lee–Tarver model was used, which allows describing both initiation and stationary propagation of detonation.

The Lee–Tarver model was first published in [31] and a modified version was described in [32]. The model uses two Johns-Wilkins-Lee

(JWL)-type EOS. The first one is used to describe the initial explosive and the second one to describe the products of the explosion:

$$p = Ae^{-R_1 V} + Be^{-R_2 V} + \omega C_V T / V$$

where p is the pressure, V is the specific volume, T is the temperature, ω is the Gruneisen coefficient, C_V is specific heat capacity, and A, B, R_1, and R_2 are constants.

The kinetic relationship itself is written in the following form [32]:

$$\frac{dF}{dt} = I(1-F)^b \left(\frac{\rho}{\rho_0} - 1 - a\right)^x + G_1(1-F)^c F^d p^y + G_2(1-F)^e F^g p^z,$$

where F is the fraction of the reacted explosive, t is time, ρ is the current density, ρ_0 is the initial density, and I, G_1, G_2, a, b, c, d, e, g, x, y, and z are constants.

This trinomial law describes the three stages of the reaction observed during shock wave initiation and detonation of heterogeneous explosives.

The first term describes the ignition of an explosive upon its compression, i.e. the creation of "hot spots." This term varies in the range $0 < F < F_{I\max}$. The second term describes the slow growth of isolated hot regions and varies within the limits $0 < F < F_{G_1\max}$. The third term describes the rapid termination of the reaction after the coalescence of hot regions. This term varies between $F_{G_2\max} < F < 1$.

The first term uses the value of compression of the initial explosive to start the combustion process a. This was done so that the model could be used to calculate isentropic and multiple shock compression with the same success as a single shock compression.

The parameters of the kinetic relationships were chosen by comparing the experimental data with the results of a series of preliminary calculations in which not only did the parameters themselves vary, but also the different EOS of unreacted explosives and detonation products were tested. The best results were obtained using the coefficients of the JWL EOS for Composition B from [33] and the corresponding kinetic model from [34].

At the beginning of the simulation, the initial detonation wave was set at the left boundary of the active charge. The conditions on the free boundaries were defined as "flow out" according to the AUTODYN terminology. The values of the parameters of governing equations for cast TG-40 are given in Table 2.10, for silicon carbide – in Table 2.4.

Details of the process in the case of a 35 mm silicon carbide insert are shown in Figure 2.18. At the time t = 5.25 μs, the detonation front approaches the end of the active charge of the explosive. The pressure at

Table 2.10 Parameters for cast TG-40.

Unreacted JWL	Product JWL
$A = 1479$ Mbar	$A = 5.308$ Mbar
$B = -0.05261$ Mbar	$B = 0.0783$ Mbar
$R_{1u} = 12$	$R_1 = 4.5$
$R_{2u} = 12$	$R_2 = 1.2$
$\omega_u = 0.912$	$\Omega = 0.34$
$C_v = 2.487 \times 10^{-5}$ Mbar/K	$C_v = 1.0 \times 10^{-5}$ Mbar/K
$T_o = 298$ K	C-J Energy/unit volume $E_{og} = 0.081$ Mbar
Shear modulus = 0.035 Mbar	C-J Detonation velocity $U_D = 7.576$ mm/μs
Yield strength = 0.002 Mbar	C-J Pressure $P_{CJ} = 0.265$ Mbar
$\rho_0 = 1.63$ g/cm^3	Reaction zone width $W_{reac} = 2.5$
Von Neumann spike rel vol. $c_0 = 0.7$	Max change in reaction ratio $\Delta F_{max} = 0.1$
C-J Energy/unit volume $E_{0,u} = -0.00504$ Mbar	
Reaction rates	
$a = 0.0367$	$x = 7.0$
$b = 0.667$	$y = 2.0$
$c = 0.667$	$z = 3.0$
$d = 0.333$	$F_{Imax} = 0.022$
$e = 0.222$	$F_{G_1\,max} = 0.7$
$g = 1.0$	$F_{G_2\,max} = 0.0$
$I = 4.0 \times 10^6$ μs^{-1}	$G_1 = 140$ Mbar^{-2}/μs
Maximum relative volume in tension = 1.1	$G_2 = 1000$ Mbar^{-3}/μs

Source: From Balagansky and Stepanov [22]. Reprinted with permission of Springer Nature.

the front of the detonation wave in the active charge is 22.1 GPa. This wave, when entering a rod of silicon carbide, creates a wave with a front pressure of 28.1 GPa. At the same time, the shock wave generated in water has an initial pressure at the front of 19.1 GPa. Subsequently, the waves in the rod and in the water propagate toward the face of the passive charge. Obviously, the wave in the rod outruns the wave in the water. By the time $t = 8.50$ μs, the wave in the rod reaches the face of the passive charge.

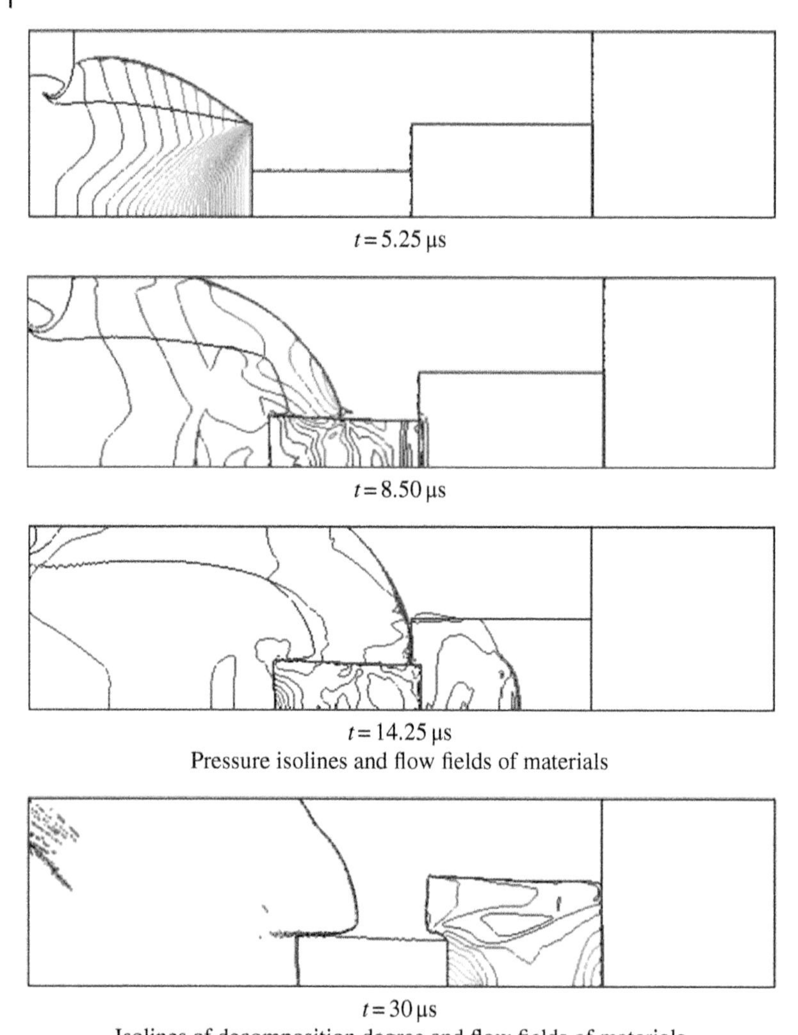

Figure 2.18 Process details for the setup with a 35 mm silicon carbide insert. (*Source:* From Balagansky and Stepanov [22]. Reprinted with permission of Springer Nature.)

It has an amplitude of 3.6 GPa and creates a compression wave with an amplitude of about 2 GPa in the passive charge; the decomposition degree of the passive charge is equal to 0. At the time $t = 14.25\,\mu s$, the shock wave in the water reaches the face of the passive charge and creates a wave with an amplitude of 2 GPa at the periphery of the flat face of the passive charge. The maximum pressure on the axis of symmetry of the passive charge is 4.6 GPa, and the maximum decomposition degree of the passive charge reaches a value of 0.2 in the zone adjacent to the boundary with the ceramic rod.

The initiation of the passive charge on the symmetry axis by a compression wave does not develop. The initiation of the charge by a wave that passed through water also does not happen, since this wave propagates through a charge that was compressed by the outrunning wave. Note that the wave that passed through water converges to the axis of symmetry. In this case, it is possible to form a Mach configuration on the symmetry axis.

The final part of Figure 2.18 shows the development of the process at time t = 30 μs. At this instance of time, the highest pressure in the passive charge does not exceed 0.8 GPa, and the maximum degree of decomposition of the passive charge is observed at the region near the end of the ceramic rod and it has practically not changed since the time of 14 μs. It should be noted that when the reflection from the surface of a steel specimen occurs, the decomposition of the passive charge occurs in the reflected wave, and to this instant, the degree of decomposition of explosives near the contact boundary reaches 0.15. In case of a detonation initiation, the process develops after the arrival of the initiating wave according to the well-known scenario.

Table 2.11 presents the comparison of calculated and experimental data. It should be noted that in the calculations, the polytetrafluoroethylene (PTFE) was replaced with polymethyl methacrylate (PMMA). A completely satisfactory agreement of the calculated and experimental results is observed, especially if one considers that the experimental determination of the values of distances is largely approximate.

From the data presented, it follows that in the scheme of the GAP test under consideration, the results can be distorted due to the initiation of the passive charge with a wave reflected from the identification specimen. In addition, one may note that the formation of Mach configurations on the axis of symmetry of the passive charge is possible as a result of the development of the process. This fact was actually confirmed in [35].

2.3 The Phenomenon of Energy Focusing in Passive High Explosive Charges

The processes that we consider in this monograph can lead to the formation of a Mach wave configuration on the symmetry axis of the passive charge of explosives. In experiments described in [35–39], this has led to

Table 2.11 Detonation transmission distances (experimental and calculated).

Inert insert material	Reliable detonation transmission distance, mm		Failure distance, mm	
	Calculated	Experimental	Calculated	Experimental
No inert insert	45	50	50	55
Polytetrafluoroethylene (PTFE)/polymethyl methacrylate (PMMA)	45	45	50	50
Copper	37	30	40	37
Silicon carbide	30	26	35	30

Source: From Balagansky and Stepanov [22]. Reprinted with permission of Springer Nature.

the appearance of an unexpected effect of energy cumulation. The effect was observed during investigation of the identification specimens that were part of the experimental assembly (see Figure 2.16). Let us dwell in more detail on the results of analysis of these samples.

2.3.1 Characterization of Steel Specimens Deformed in Experiments on Energy Focusing

As mentioned earlier, the development of detonation or its absence in the experiments was judged by visual control of the specimens. When detonation developed in the passive charge, a characteristic depression can be found on the surface of the specimen. In some of the experiments, a crater with a depth of about 10 mm and a diameter of about 5 mm was observed in the center of the depression. Such a crater obtained in an experiment with a ceramic insert with length of 19.5 mm is shown in Figure 2.19. In the area contacting with the charge, the surface has typical temper colors, which indicates the local effect of high temperatures. A similar phenomenon was also observed in experiments with copper inserts, but it was manifested to a lower degree. The surface of the identification specimen for experiments with the copper insert is shown in Figure 2.20. In experiments without inserts or with PTFE inserts that led to a detonation transmission, only a depression without any specific features was observed on the surface of the specimens (Figure 2.21).

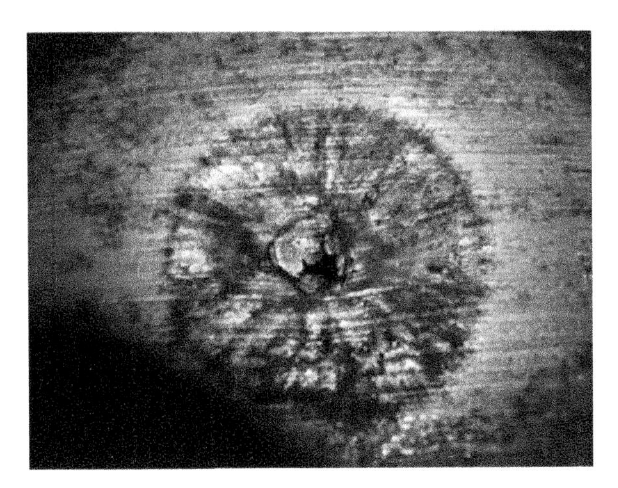

Figure 2.19 Surface of the identification specimen with a crater. (*Source:* From Balagansky et al. [30]. Reprinted with permission of World scientific Publishing Company.)

Figure 2.20 Surface of the identification specimen for a setup with a copper insert.

Figure 2.21 Typical surface of the identification specimen for experiments without inserts and with polytetrafluoroethylene (PTFE) inserts.

During the experiments, the inert inserts made of ceramics and PTFE were completely disintegrated, while the copper inserts were retained in a deformed state. Visual control of the copper inserts recovered after the experiments showed that inserts having initial length of 20 mm had indentations (sometimes even through holes) on the surface facing the passive charge. Apparently, this is a consequence of the cumulation of the detonation products of the passive charge in the direction opposite to the direction of the detonation transmission. A typical copper insert illustrating this effect is shown in Figure 2.22.

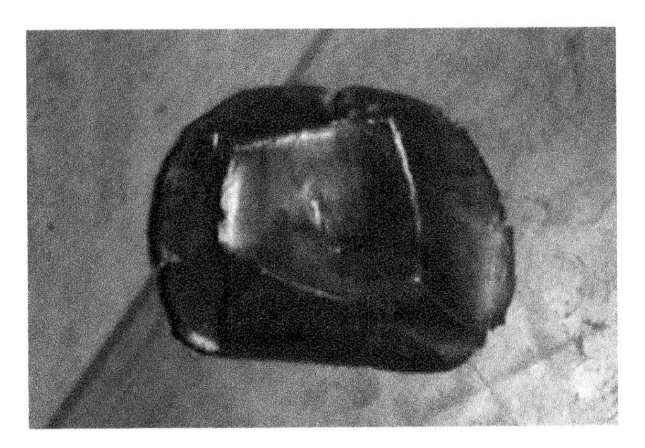

Figure 2.22 Deformed copper insert. (*Source:* From Balagansky et al. [30]. Reprinted with permission of World scientific Publishing Company.)

Figure 2.23 Deformation twins formed in laminated pearlite colonies of the identification specimen (optical microscope). (*Source:* From Balagansky et al. [35]. Reprinted with permission of AIP Publishing.)

In the initial state, the specimens had a ferrite–pearlite structure with an average ferrite grain size of 28 μm. Pearlite had a lamellar structure and was observed as relatively equiaxed conglomerates of several colonies. After explosive loading, a significant amount of deformation twinning was formed not only in the ferrite grains but also in colonies of pearlite (Figures 2.23 and 2.24). Such structural changes are rarely observed for classical methods of plastic deformation and even for most methods of high-strain-rate deformation. These structural changes indirectly indicate high pressures and temperatures occurring

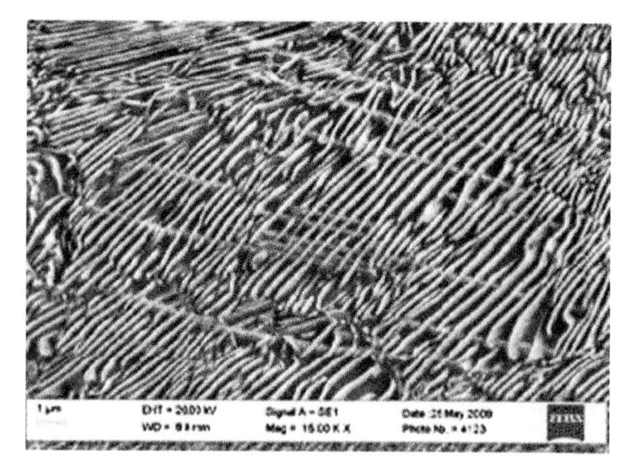

Figure 2.24 Deformation twins formed in laminated pearlite colonies of the identification specimen (scanning electron microscope). (*Source:* From Balagansky et al. [35]. Reprinted with permission of AIP Publishing.)

Figure 2.25 A surface of a steel identification specimen for ceramic insert with height of 16.5 mm. (*Source:* From Balagansky et al. [35]. Reprinted with permission of AIP Publishing.)

in this explosive loading scheme and confirm the conclusions of previous sections [39].

This scheme of loading was used in several additional experiments; however, the height of the ceramic inserts varied (16.5, 18, and 20 mm). The experiments showed stable reproducibility of the described effects. The formation of crater was most pronounced with inserts of 16.5 and 18 mm in height. Photographs of the corresponding identification specimens are shown in Figures 2.25 and 2.26.

Figure 2.26 A surface of a steel identification specimen for ceramic insert with height of 18.0 mm.

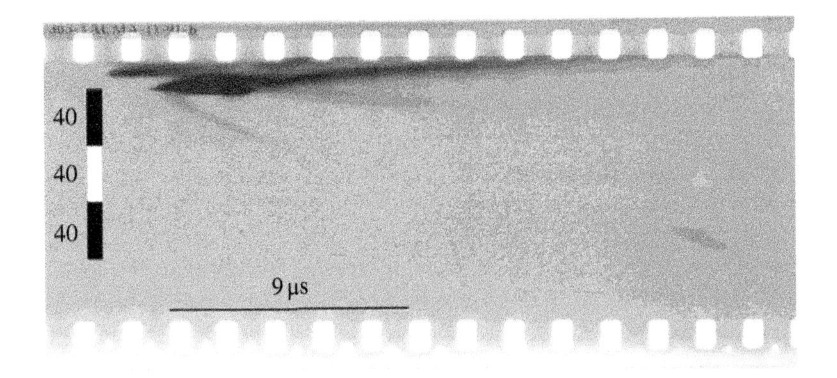

Figure 2.27 Streak camera photograph for experiment without inert insert with 40 mm gap.

2.3.2 Optical Recording in Streak Mode

The peculiarities of the processes occurring in the described explosive systems were investigated by optical recording using VFU high-speed camera in streak mode. The angular rotation speed of the mirror in all experiments was 60 000 rpm. The scale bars of time (μs) and length (mm) are shown in Figures 2.27–2.29. Certain distortions in the streak images of wave fronts due to the refraction of light in a glass of water obstruct accurate determination of detonation velocities in the active and passive charges. At the same time, it is possible to determine with sufficient accuracy the time from the moment when the detonation wave exits the active charge to the instant when initiation of the passive explosive charge starts. This data became the main experimental information obtained in this series of experiments.

The sequence of streak images for the experiment without an inert insert with 40 mm gap between the active and passive charges is shown in Figure 2.27. One can relatively easily estimate that detonation in a passive charge comes to its lateral surface and thus can be detected at a distance equal to approximately half the length of the charge, i.e. at a distance of 20 mm from the end. The total time elapsed from the moment when detonation wave comes to the end of the active charge until the emergence of the detonation wave on the lateral surface of the passive charge is approximately equal to 14.6 μs. The calculated time, which takes the shock wave to pass a distance of 40 mm in water between the active and passive charges, is approximately equal to 10.0 μs. Thus, the delay from the moment of the shock wave arrival to the end face of the passive charge until the moment of detonation development in the passive charge and its exit to the lateral surface is approximately 4.6 μs.

A similar sequence of images for the experiment with a 30 mm inert copper insert is shown in Figure 2.28. In this case, the total time elapsed from the moment when the detonation wave comes to the end of the active charge before the emergence of the detonation wave on the lateral surface of the passive charge is approximately 18.0 μs. According to the high-speed camera data, the appearance of detonation wave on the lateral surface of the passive charge occurs already in the region adjacent to the far end of the charge. It may be noticed that the detonation moves both toward the far end of the passive charge and in the opposite direction. The calculated time, which takes the shock wave in the water to pass the distance between the active and passive charges, is about 6.6 μs. The estimated time for which the shock wave passes the distance from the

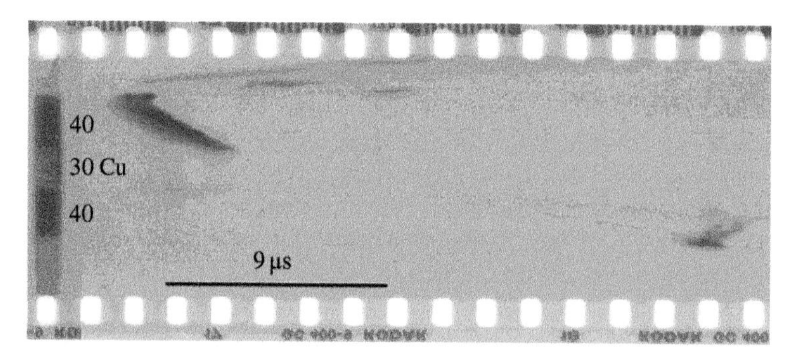

Figure 2.28 Streak camera photograph for experiment with 30 mm copper insert.

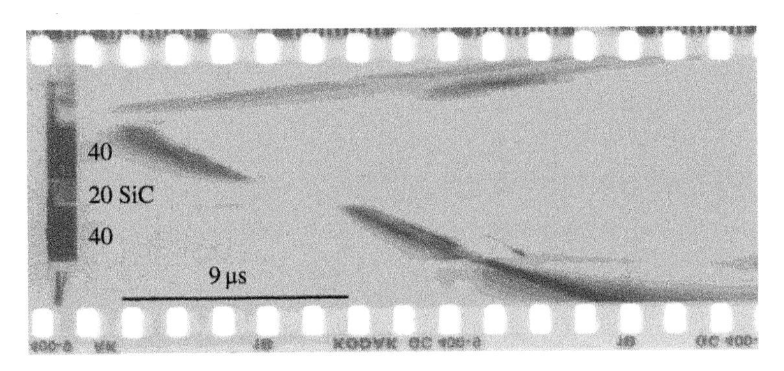

Figure 2.29 Streak camera photograph for the experiment with a 20 mm silicon carbide insert.

active to the passive charge in the copper rod is approximately 6.4 μs. Thus, the delay from the arrival of the shock wave in the water to the end of the passive charge until the moment of detonation development in the passive charge and its appearance on the lateral surface is approximately 11.4–11.6 μs.

In Figure 2.29, a sequence of streak images for the experiment with a 19.6 mm highly inert insert of silicon carbide is shown.

In this experiment, the steel identification specimen was moved away from the rear end of the passive charge by 50 mm to control the cumulation of the detonation products. In this case, the total time elapsed from the moment when detonation wave comes to the rear face of the active charge until the emergence of the detonation wave on the lateral surface of the passive charge is approximately equal to 4.0 μs. The time that takes the shock wave to pass the distance between the active and passive charges in water was estimated to be approximately equal to 4.0 μs. The time for an elastic wave to pass a distance of 20 mm in a rod of silicon carbide is approximately equal to 1.8 μs. The amplitude of this wave is 2.2 GPa; the front is blurred by approximately 3.3 mm.

This wave does not excite the detonation in a passive charge. However, in this case, there is almost no delay in initiating the detonation in a passive charge by a shock wave in water.

The analysis of the streak mode observations of the detonation transfer process for various conditions indicates that when detonation is transmitted at distances close to the limiting values, a significant delay in the excitation of detonation in the passive charge is observed.

In the experiment with a 19.6 mm long insert made of silicon carbide and an identification specimen moved forward by 50 mm, the detonation products fly apart at a rate approximately equal to the detonation velocity of the explosive of a passive charge. Apparently, these products, focusing to the axis of symmetry, create a crater in the specimen.

2.3.3 Optical Recording in Frame Mode

In the first series of experiments with cast TG-40 in the frame mode, the detonation process was recorded using a high-speed VFU camera with the mirror speed of 53 000 rpm, which provided a time interval between frames of 4.53 μs. The images were recorded using a background with a reference grid having a cell size of 15 × 15 mm. As opposed to the experiments with the streak mode recording, the water container was made of glass and had flat walls to eliminate optical distortions. Shadow explosive argon illumination was used.

The typical sequence of frames is shown in Figure 2.30. On the last frames, one may observe noticeable heterogeneities appearing at the rear end of the passive charge. This means that instability develops on the contact boundary of media with different densities, i.e. between detonation products and water. This was a general property of this process, which was established by the analysis of data obtained by high-speed optical photography of the wave propagation process in experimental setups in both streak and frame modes [38]. It can be asserted that when the wave leaves the rear end of the passive charge, at the end of the charge, local inhomogeneities are formed. Apparently, this is due to the development of the Rayleigh–Taylor instability at the interface of two media or due to the instability of the detonation front in the passive charge itself. The front of the wave exiting water is convex in the direction of motion, indicating that the velocity on the symmetry axis is significantly higher than at the periphery of the charge. It is logical to assume that this effect is associated with the initiation of detonation in a preliminary compressed explosive charge and the convergence of the detonation wave to the axis of symmetry of the charge.

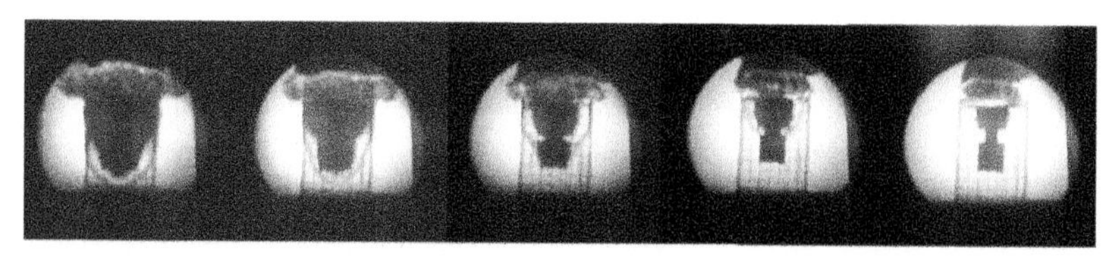

Figure 2.30 Frames of the process of detonation transmission through water with a 19.6 mm silicon carbide insert. Consecutive frames from right to left in 4.53 μs increments. (*Source:* From Balagansky et al. [35]. Reprinted with permission of AIP Publishing.)

In the second series of experiments, Composition B and SEP explosives were used. Composition B is similar in composition and properties to TG-40. SEP is a plastic explosive based on pentaerythritol tetranitrate (PETN). When setting up the experiments, we tried to prevent the deviation of components from the axis of symmetry, to avoid the gaps between the explosives and the supporting elements (to avoid the channel effect) and to eliminate the presence of air bubbles in the water. These factors, as established in preliminary experiments, significantly affect the result.

The results of all experiments with silicon carbide inserts and identification specimen are given in Table 2.12, where the last column indicates the presence or absence of cumulation of the explosion products.

Recordings with the high-speed video camera SHIMADZU HPV-1 were conducted for ceramic insert heights of 15, 16, 17, 18, and 20 mm and cast Composition B with density of 1.66–1.67 g/cm^3. For those experiments, videos and still frames with a time step of 1 μs were recorded. A high-speed recording was made for experimental assemblies without identification specimen under back-surface impulse lighting. In Figure 2.31, we can see the plane wave generator, active and passive HE charges, 20 mm ceramic insert, and strips fastened together with Scotch tape.

The sequence of images recorded in experiments with ceramic inserts of 18 and 20 mm in height is shown in Figure 2.32. The moments of time, counted from the time when the detonation wave exits to the end of the active charge, are indicated in the photographs. Despite the difference in

Table 2.12 Focusing phenomenon presence.

HE, density	Height of insert Δ, mm	Number of shots	Result (yes/no)
Cast TG-40, 1.65 g/cm^3 (Russian analog of Composition B)	16.5	1	Yes
	18.0	1	Yes
	19.5	3	Yes
	20.0	2	Yes
	24.8	1	No
Cast Composition B, 1.66–1.67 g/cm^3	18.0	1	Yes
	20.0	1	No
SEP, 1.30 g/cm^3 (pentaerythritol tetranitrate [PETN]-based plastic HE)	10.0	1	No
	15.0	1	Yes
	20.0	1	Yes

Source: From Balagansky et al. [35]. Reprinted with permission of AIP Publishing.

Figure 2.31 Assembly with an 18 mm ceramic insert height. (*Source:* From Balagansky et al. [35]. Reprinted with permission of AIP Publishing.)

the height of the inserts of only 2 mm, the processes in the experiments differ not only quantitatively but also qualitatively. The process in the experiment with 20 mm long inserts lags behind the process with 18 mm long inserts not only due to the higher insert height but also due to the slower development of detonation in the passive charge. When the detonation wave reaches the end of the passive charge, an accelerated motion of the detonation products from the end of the passive charge in the water is observed.

Analysis of the photographs indicates that the visible glow corresponds to the detonation front inside the explosive charge, since the front of the shock wave in water is in some cases noticeably behind the glowing front in the charge.

Particular attention should be paid to the third and fourth frames in the right column of Figure 2.32, which clearly shows the formation of a Mach configuration of detonation waves.

An analysis of the entire set of obtained data allows to formulate the following physical model of the observed phenomena. The outrunning compression wave in the ceramic insert enters the passive charge of the explosive, resulting in desensitization of the explosive. Further, a shock wave in water comes to the peripheral region of the front end of the passive charge, initiating a detonation wave converging to the axis of symmetry. This greatly increases the velocity and pressure at the front and forms the Mach wave on the axis of the precompressed passive charge. It should be noted that the process develops when conditions of the passive charge initiation are close to critical ones.

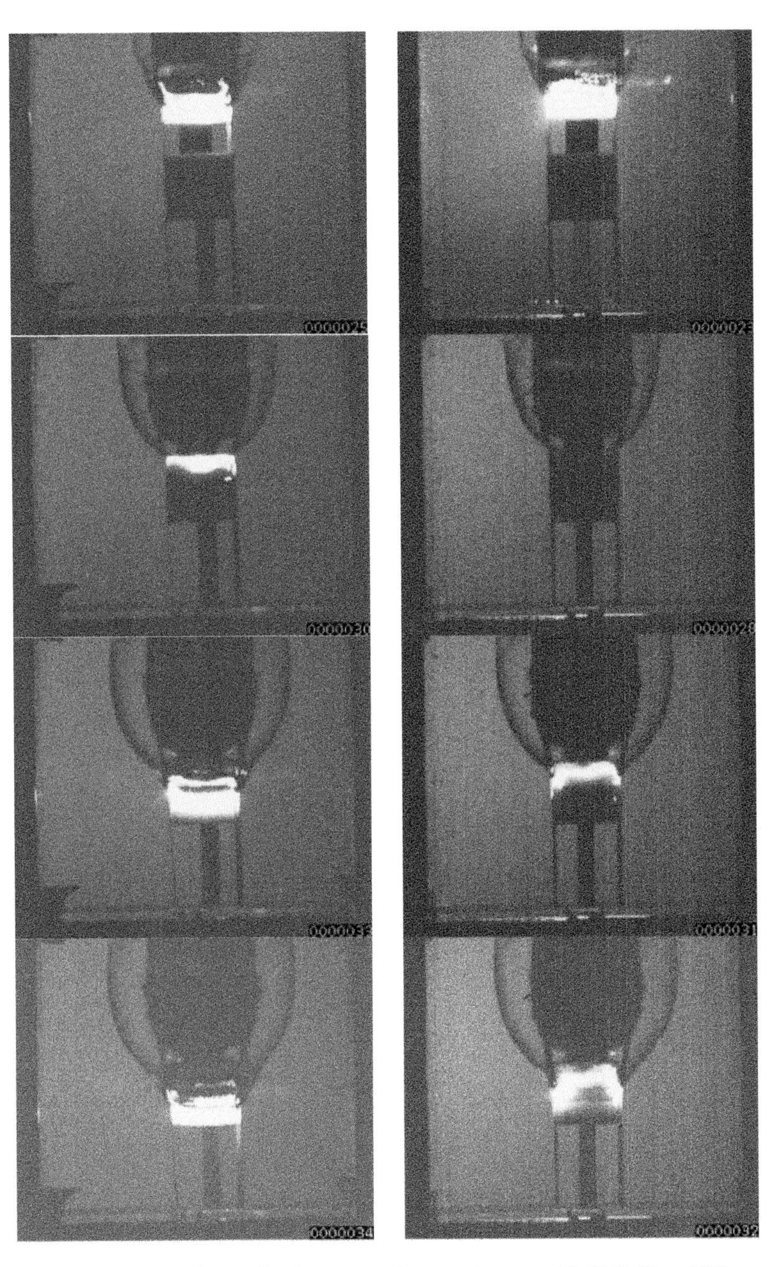

Figure 2.32 Optical records of processes in experiments with 18 (left) and 20 mm (right) ceramic inserts. Time step is equal to 1 μs. (*Source:* From Balagansky et al. [35]. Reprinted with permission of AIP Publishing.)

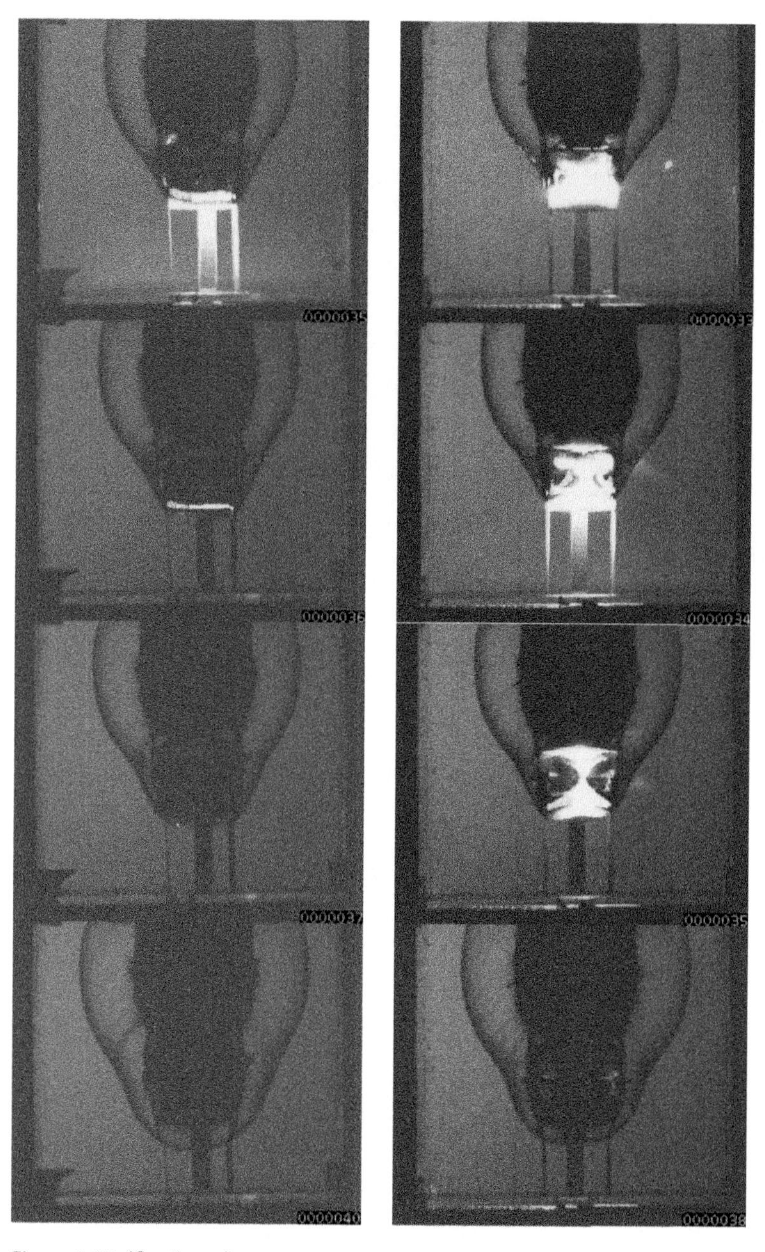

Figure 2.32 (Continued)

A similar phenomenon of the concentration of energy in converging shock waves was observed, for example, in experiments on initiation of detonation in gaseous explosives by a fast-flying, ring-shaped body or impact of ring hammers against massive barriers [40–42].

2.3.4 Numerical Modeling of the Energy Focusing Phenomenon

Numerical simulation of the processes described in [35] was performed in a 2D-axis symmetry formulation on an Eulerian fixed mesh with a spatial resolution of 2600 × 800 cells (Figure 2.33) using the ANSYS AUTODYN package, which was described earlier (see also Appendix B).

Initial posting corresponded to the experimental assembly for $\Delta = 20\,\text{mm}$ (see Figure 2.16). The boundary conditions on the left, right, and upper boundaries were defined as "flow out" in AUTODYN terminology. Constants for air, steel, and water were selected from the standard AUTODYN EOS library. As before (see Section 2.2.2), the behavior of a passive charge with a density of $1.63\,\text{g/cm}^3$ was calculated using JWL EOS and the Lee–Tarver kinetics. A detonation wave with a plane front was initiated in the active charge of the explosive and propagated from left to right. The results of the calculation are shown in Figure 2.34.

Figure 2.34 shows material flow fields and pressure contours at consecutive moments of time from the beginning of the process. The calculation illustrates the features of the process leading to formation of the Mach configuration of detonation waves.

Figure 2.35 presents a graph of the pressure distribution along the symmetry axis at time $t = 15.75\,\mu\text{s}$. The maximum pressure at the front of the Mach wave is 1.25 Mbar.

It should be noted that the Mach wave occurs in TG-40 over a wider range of conditions than in Composition B. This may be due to differences in the microstructure of charges with the same chemical composition. The best conditions for Mach wave formation were achieved using TG-40 and ceramic inserts with lengths of 16.5, 18.0, or 19.5 mm.

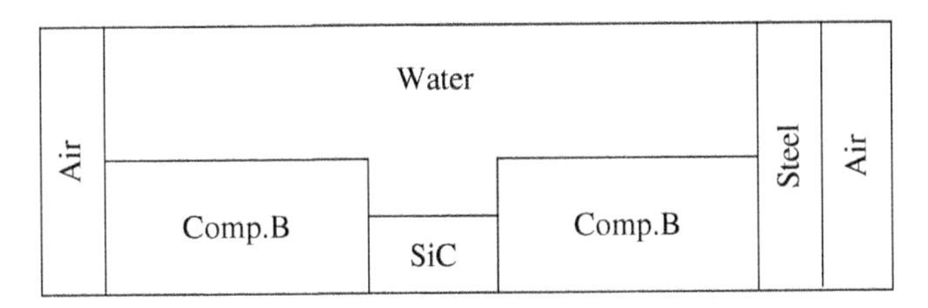

Figure 2.33 Initial posting with dimensions of 13.0 cm × 4.0 cm. The bottom boundary is the symmetry axis. The direction of wave propagation is from left to right. (*Source:* From Balagansky et al. [35]. Reprinted with permission of AIP Publishing.)

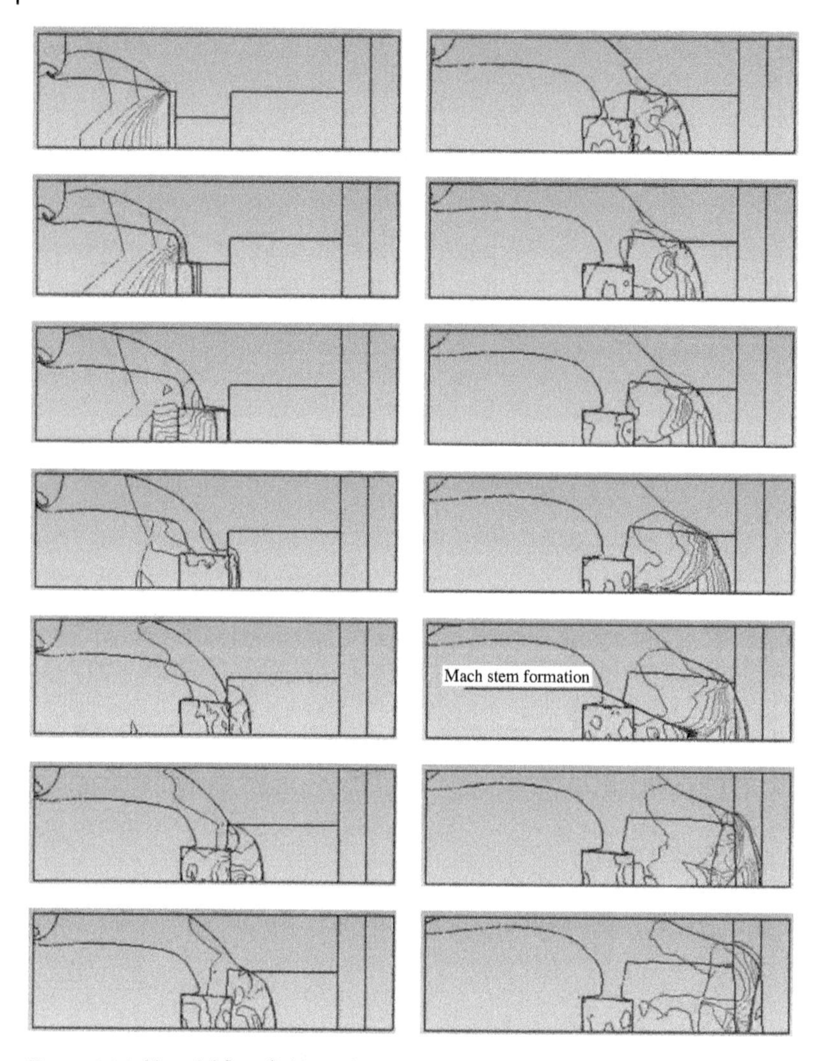

Figure 2.34 Material flow fields and pressure contours at different times with time step of 1 μs. (*Source:* From Balagansky et al. [35]. Reprinted with permission of AIP Publishing.)

2.4 Summary

The results of the experiments indicate that a compression wave with a strong smoothening of its front propagates along the ceramic rod with the elastic wave velocity. This is consistent with both the theoretical concept of elastic shock wave smoothening in the body of finite transverse

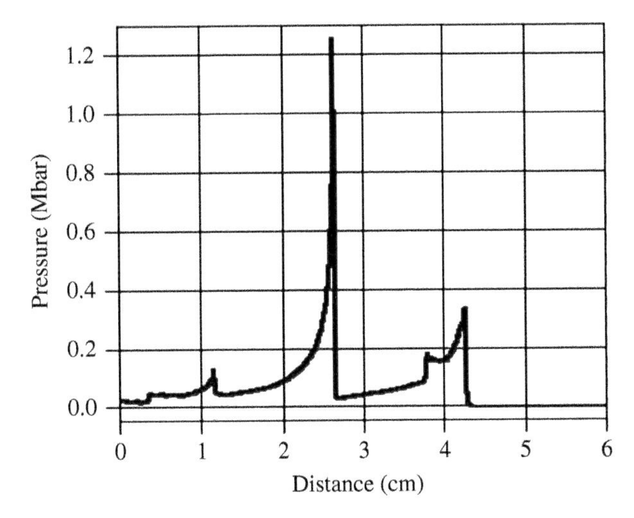

Figure 2.35 Pressure graph in the passive HE charge versus axial distance at $t = 15.75\,\mu s$ from $x_1 = 7\,cm$ to $x_2 = 13\,cm$. (*Source:* From Balagansky et al. [35]. Reprinted with permission of AIP Publishing.)

dimensions and the corresponding numerical estimations. The evolution of shock waves, which was observed in experiments, is quite adequately described in the framework of the numerical simulations, which makes it possible to analyze the details of the processes.

The passive charge loading by an outrunning wave through copper and ceramic rods leads to significant desensitization of the passive charge of TG-40. It should be noted that a compression wave with a diffuse front propagating along the ceramic rod has even greater desensitizing ability. As a result of the process development, the formation of Mach configurations on the symmetry axis of the passive charge is possible, which was evidenced by the analysis of the identification specimen surface (the presence of cavities and microstructural changes). This is also confirmed by an analysis of streak mode images of the detonation transmission process.

An analysis of the whole data obtained in experiments and simulations leads to the following formulation of a physical model of the observed phenomena. The outrunning compression wave in the ceramic insert enters the passive charge of explosive, resulting in its desensitization. Then the shock wave propagated though the water comes to the peripheral region of the front end of the passive charge, initiating the detonation wave converging to the axis of symmetry. This greatly increases the velocity and pressure at the front and forms a Mach wave on symmetry axis of the precompressed passive charge of explosive. It should be noted that the process develops when conditions of a passive charge initiation are close to critical ones.

References

1 Merzhievsky, L.A. (1982). *Shock Waves in Condensed Matter: Training Manual.* Novosibirsk: Novosibirsk State University [in Russian].

2 Kanel, G.I. and Molodets, A.M. (1976). Behavior of type K-8 glass under in dynamic compression and subsequent unloading. *Zhurnal Tekhnicheskoi Fiziki* **46** (2): 398–407. [in Russian].

3 Grady, D.E. (1994). Shock-wave strength properties of boron carbide and silicon carbide. *Journal de Physique IV* **4** (C8): 385–391.

4 Rayleigh, J.W.S. (1894). *The Theory of Sound.* London: Macmillan.

5 Love, A.E.H. (1906). *A Treatise on the Mathematical Theory of Elasticity.* Cambridge: University Press.

6 Rabotnov, Y.N. (1979). *Mechanics of Deformable Solids.* Moscow: Nauka [in Russian].

7 Merzhievskii, L.A. (1987). Front width of a stationary shock wave in a metal. *Combustion, Explosion and Shock Waves* **23** (3): 370–372.

8 Balagansky, I.A., Balagansky, A.I., Razorenov, S.V. et al. (2005). Evolution of shock waves in silicon carbide rods. *Proceedings of the International Conference 'VII Khariton Topical Scientific Readings'*, Sarov, Russia (14–18 March 2005). Sarov: RFNC. [in Russian].

9 Balagansky, I.A., Balagansky, A.I., Razorenov, S.V. et al. (2006). Evolution of shock waves in silicon carbide rods. *Proceedings of the 14th APS Topical Conference on Shock Compression of Condensed Matter*, Baltimore, USA (31 July–5 August 2005). Melville: AIP Publishing.

10 Balagansky, I.A., Balagansky, A.I., Razorenov, S.V. et al. (2007). Evolution of shock waves in silicon carbide rods. *Proceedings of the III Scientific and Technical Conference 'Actual Problems and Prospects for the Development of Low-Sensitivity Energy Materials and Low-Risk Products'*, Dzerzhinsk, Russia (23–25 June 2004). Dzerzhinsk: GosNII Kristall. [in Russian].

11 Brar, N.S. and Bless, S.J. (1992). Dynamic fracture and failure mechanisms of ceramic bars. In: *Shock Wave and High-Strain-Rate Phenomena in Materials* (ed. M.A. Meyers, L.E. Murr and K.P. Staudhammer), 1041–1049. New York: Marcel Dekker.

12 Century Dynamics (2005). *Autodyn. Explicit Software for Nonlinear Dynamics.* Theory manual.

13 Wilkins, M.L. (1964). Calculation of elastic-plastic flows. In: *Methods in Computational Physics*, vol. **3** (ed. B. Alder, S. Fernbach and M. Rotenberg), 211–263. New York/London: Academic Press.

14 Jacobs, S.J. (1960). Non-steady detonation. *Proceedings of 3rd Symposium on Detonation*, Prinston, USA (26–28 September 1960). Arlington: Office of Naval Research.

15 Campbell, A.W., Davis, W.C., Ramsay, J.B. et al. (1961). Shock initiation of solid explosives. *Physics of Fluids* **4**: 511–521.

16 Campbell, A.W. and Travis, J.R. (1985). The shock desensitization of PBX-9404 and Composition B-3. *Proceedings of 8th Symposium on Detonation*, Albuquerque, USA (15–19 July 1985). Silver Spring: Naval Surface Weapons Center.

17 Davis, W.C. (2010). Shock desensitizing of solid explosive. *Proceedings of 14th International Detonation Symposium*, Coeur d'Alene, USA (11–16 April 2010). Arlington: Office of Naval Research.

18 Tarzhanov, V.I. (1976). Detonation velocity of shock-compressed cast TNT. *Combustion Explosion and Shock Waves* **12**: 810–814.

19 Bordzilovskii, S.A. and Karakhanov, S.M. (1995). Desensitization of pressed RDX/paraffin and HMX/paraffin compounds by multiple shock waves. *Combustion, Explosion and Shock Waves* **31**: 227–235.

20 Setchell, R.E. (1981). Ramp wave initiation of granular explosives. *Combustion and Flame* **43**: 255–264.

21 Setchell, R.E. (1983). Effects of precursor waves in shock initiation of granular explosives. *Combustion and Flame* **54**: 171–182.

22 Balagansky, I.A. and Stepanov, A.A. (2016). Numerical simulation of Composition B high explosive charge desensitization in gap test assembly after loading by precursor wave. *Shock Waves* **26**: 109–115.

23 Balagansky, I.A. and Gryaznov, E.F. (1994). Desensitization of RDX-charges after preshocking by compression wave in SiC-ceramic rod. *Proceedings of International Conference on Combustion 'Zel'dovich Memorial'*, Moscow, Russia (12–17 September 1994). Moscow: Russian Section of the Combustion Institute.

24 Baum, F.A., Stanyukovich, K.P., and Shekhter, B.I. (1959). *Physics of the Explosion*. Moscow: Fizmatgiz [in Russian].

25 Balagansky, I.A., Matrosov, A.D., Stadnichenko, I.A. et al. (2007). Influence of inert copper and silicon carbide inserts on process of detonation transmission through water. *Proceedings of the International Conference 'IX Khariton Topical Scientific Readings'*, Sarov, Russia (12–16 March 2007). Sarov: RFNC. [in Russian].

26 Balagansky, I.A., Matrosov, A.D., Stadnichenko, I.A. et al. (2007). Behavior of condensed heterogeneous high explosive under shock wave loading through water after preloading with advanced wave. *Doklady Akademii Nauk Vyssej Skoly Rossii* **2** (9): 76–83. [in Russian].

27 Balagansky, I.A., Matrosov, A.D. Stadnichenko, I.A. et al. (2007). Influence of inert copper and silicon carbide inserts on process of detonation transmission through water. *Program and Abstract of 2nd International Symposium on Explosion, Shock Wave and Hypervelocity Phenomena*, Kumamoto, Japan (6–9 March 2007). Kumamoto: Kumamoto University.

28 Balagansky, I.A., Matrosov, A.D., Stadnichenko, I.A. et al. (2007). Desensitization of heterogeneous high explosives under initiation through high modulus elastic elements. *Abstracts Book of Sixth International Symposium on Impact Engineering*, Daejeon, Korea (16–19 September 2007). Daejeon: ISIE2007 Organizing Committee.

29 Balagansky, I.A., Matrosov, A.D., Stadnichenko, I.A. et al. (2008). Influence of inert copper and silicon carbide inserts on process of detonation transmission through water. *Materials Science Forum* **566**: 207–212.

30 Balagansky, I.A., Matrosov, A.D., Stadnichenko, I.A. et al. (2008). Desensitization of heterogeneous high explosives under initiation through high modulus elastic elements. *International Journal of Modern Physics B* **22**: 1305–1310.

31 Lee, E.L. and Tarver, C.M. (1980). Phenomenological model of shock initiation in heterogeneous explosives. *Physics of Fluids* **23** (12): 2362–2372.

32 Tarver, C.M., Hallquist, J.O., and Erickson, L.M. (1985). Modeling short pulse duration shock initiation of solid explosives. *Proceedings of 8th Symposium on Detonation*, Albuquerque, USA (15–19 July 1985). Silver Spring: Naval Surface Weapons Center.

33 Murphy, M.J., Lee, E.L., Weston, A.M. et al. (1993). Modeling shock initiation in composition B. *Proceedings of the 10th International Detonation Symposium*. Boston, USA (12–16 July 1993). Arlington: Office of Naval Research.

34 Urtiew, P.A., Vandersall, K.S., Tarver, C.M. et al. (2006). Shock initiation experiments and modeling of composition B and C-4. *Proceedings of the 13th International Detonation Symposium*, Norfolk, USA (23–28 July 2006). Arlington: Office of Naval Research.

35 Balagansky, I.A., Hokamoto, K., Manikandan, P. et al. (2011). Mach stem formation in explosion systems, which include high modulus elastic elements. *Journal of Applied Physics* **110**: 123516.

36 Balagansky, I.A., Hokamoto, K., Manikandan, P. et al. (2009). Phenomena of energy focusing in explosive systems, which include high modulus elastic elements. *Proceedings of the 16th American Physical Society Topical Conference on Shock Compression of Condensed Matter*, Nashvill, USA (28 June–3 July 2009). Melvill: American Institute of Physics.

37 Balagansky, I.A., Hokamoto, K., Manikandan, P. et al. (2010). Study of energy focusing phenomenon in explosion systems, which include high modulus elastic elements. *Proceedings of the 14th International Detonation Symposium*, Coeur d'Alene, USA (11–16 April 2010). Arlington: Office of Naval Research.

38 Balagansky, I.A., Matrosov, A.D., Stadnichenko, I.A. et al. (2007). Cumulative phenomena in charges of condensed high explosives containing high-modulus elastic elements. *Abstracts of IX International Conference 'Zababakhin scientific readings'*, Snezhinsk, Russia (10–14 September 2007). Snezhinsk: RFNC. [in Russian].

39 Balaganskiy, I.A., Hokamoto, K., Manikandan, P. et al. (2010). Experimental study of the of energy focusing phenomenon in explosion systems, including high-modulus elastic elements. *Doklady Akademii Nauk Vyssej Skoly Rossii* **1** (14): 62–73. [in Russian].

40 Vasil'ev, A.A. and Laptev, V.I. (2014). On a high-velocity annular impactor. *Combustion, Explosion, and Shock Waves* **50**: 495–497.

41 Merzhievsky, L.A., Vasil'ev, A.A., Vinogradov A.V. et al. (2014). High-speed impact by the impactor of the ring shape. *Abstracts of XII international conference 'Zababakhin scientific readings'*, Snezhinsk, Russia (2–6 June 2014). Snezhinsk: RFNC. [in Russian].

42 Merzhievsky, L.A., Vasil'ev, A.A., Vinogradov, A.V. et al. (2015). Modeling of high-velocity impact by tubular and ring impactors. *Proceedings of the XXIV All-Russian Conference 'Numerical methods for solving problems in the theory of elasticity and plasticity'*, Omsk, Russia (2–4 June 2015). Novosibirsk: Institute of Theoretical and Applied Mechanics SB RAS [in Russian].

3

Nonstationary Detonation Processes at the Interface Between High Explosive and Inert Wall

When a detonation wave propagates along the wall of a material in which the sound velocity is greater than the detonation velocity, transient processes occur at the interface between the wall and explosive. The problem of the corresponding wave configuration near the boundary is a classical problem solved using shock polars [1, 2]. Its solution is repeatedly used in the analysis of the flows at the contact boundary between a detonating explosive and an inert medium (see some examples in [3, 4]). It may be shown by calculation that when explosive is in contact with a silicon carbide wall, the shock polar for the inert material does not intersect with the release adiabat for the explosive, i.e. the regular reflection regime is impossible, and the compression of the explosive occurs ahead of the front of the detonation wave or detonation propagates at a velocity greater than normal [5]. Thus, it was established in [6] that when detonation propagates in triaminotrinitrobenzene (TATB)-based explosive at the boundary with a beryllium plate (sound velocity of beryllium is about 13 km/s), the detonation velocity increases by 1.2%. The effect of an alumina wall was studied in [7]. It was found that the shock propagation process near the contact boundary is affected not only by the shock wave running ahead of the detonation front but also by an elastic precursor. In the subsequent sections, the results of experimental and numerical studies of processes occurring at the boundary between explosive and silicon carbide plates are presented.

Explosion Systems with Inert High-Modulus Components: Increasing the Efficiency of Blast Technologies and Their Applications, First Edition. Igor A. Balagansky, Anatoliy A. Bataev, and Ivan A. Bataev.

3.1 Measurements with Manganin Gauges

In this series of experiments, the peculiarities of the detonation wave propagation along the "explosive/silicon carbide" interface were studied with the help of manganin gauges [8–11]. The experimental assemblies for measuring the detonation parameters shown in Figure 3.1 were prepared as "sandwiches" consisting of a sample of explosive placed in contact with the polished surface of a silicon carbide plate.

The pressure profiles and the average detonation velocity on the contact surface of the explosive charge with a flat plate of silicon carbide were measured in the experiments. Several types of explosives, namely, pressed and cast trinitrotoluene (TNT), TG-40, and pressed phlegmatized hexogen (RDX) (A-IX-1), were used in the experiments. The thickness of the explosive charges was 24 mm. The silicon carbide plate had dimensions of 63 × 60 × 10 mm. The detonation wave was induced by a plane wave generator (PWG). It propagated along the polished surface of the ceramic plate. To eliminate possible gaps, the experimental setup was assembled using a vacuum lubricant. The distance from the PWG to the ceramic plate was about 10 mm.

Manganin gauges were encapsulated in fluoroplastic insulation, which had a thickness of 200 μm. They were in the shape of strips: 30–40 mm long, 0.5 mm wide, and 20 μm thick. The gauges were placed between the charge and the plate and were oriented parallel to the front of the detonation wave. In several experiments, an aluminum or copper plate (gasket) was installed between the gauges and the explosive charge. These experiments showed that the presence of a metal plate does not lead to any peculiarities of the pressure profiles.

The initial data and the results of the experiments are summarized in Table 3.1 and are represented in the graphs shown in Figures 3.2–3.7.

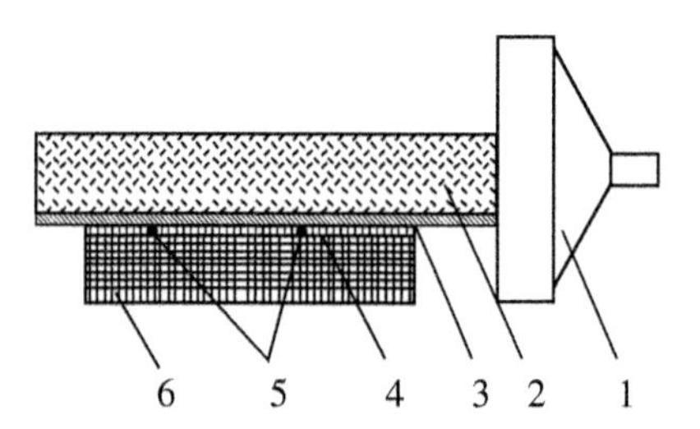

Figure 3.1 Experimental assemblies for measuring detonation parameters: 1, plane wave generator (PWG); 2, high explosive (HE); 3, metal plate; 4, gauge insulation; 5, manganin gauges; and 6, ceramic plate. (*Source:* From Balaganskii et al. [10]. Reprinted with permission of Springer Nature.)

Table 3.1 Initial data and test results.

High explosive	ρ_0, g/cm^3	Chapman–Jouguet parameters P_{CJ}, GPa	D_{CJ}, km/s	Plate material and h, mm	P_1, GPa	P_2, GPa	D, km/s
Pressed trinitrotoluene (TNT)	1.60	19.0	6.94	Cu; 1.2	14.0	16.4	7.02
				No	12.2	14.6	7.08
Cast TNT	1.60	19.0	6.94	No	18.0	16.5	6.86
Cast TNT/RDX 36/64 (TG-40)	1.67	25.4	7.80	No	17.1	17.0	7.37
Phlegmatized RDX (A-IX-1)	1.60	25.6	8.0	Al; 1.34	20.0	19.5	8.44
				No	17.0	17.8	8.95

Source: From Balaganskii et al. [10]. Reprinted with permission of Springer Nature.

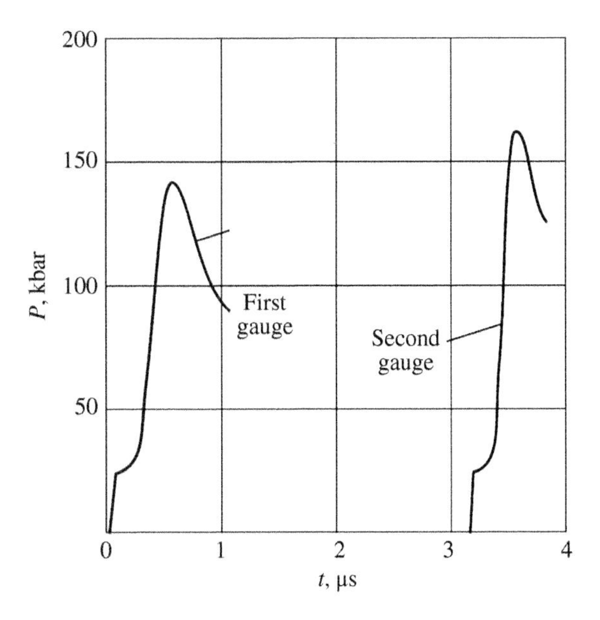

Figure 3.2 Pressure profiles recorded with manganin gauges. Pressed trinitrotoluene (TNT) with a copper gasket of 1.2 mm.

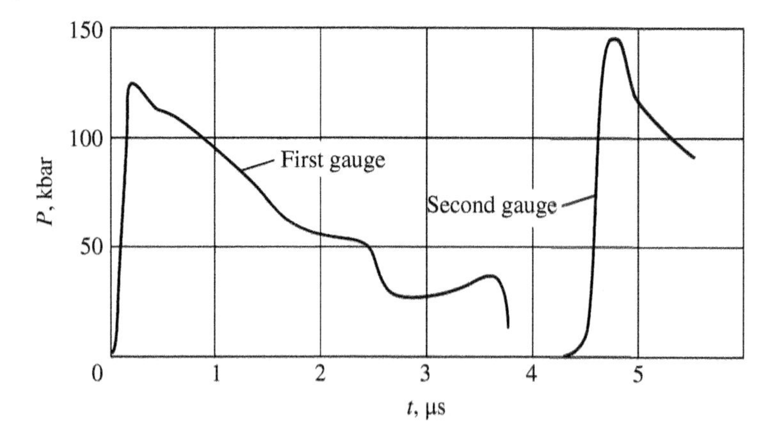

Figure 3.3 Pressure profiles recorded with manganin gauges. Pressed TNT without a gasket.

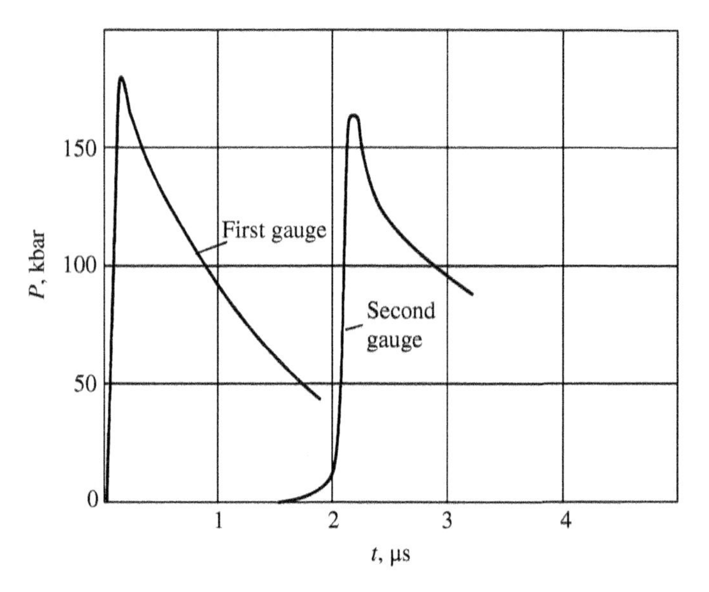

Figure 3.4 Pressure profiles recorded with manganin gauges. Cast TNT without a gasket.

The detonation velocity was calculated from the time of the pressure peaks passage measured by the first and the second gauges.

The pressure profiles for pressed TNT are shown in Figures 3.2 (with copper gasket) and 3.3 (without gasket). Similar data for experiments with cast TNT are shown in Figure 3.4, the data for experiments with TG-40 are shown in Figure 3.5, and the results for phlegmatized RDX

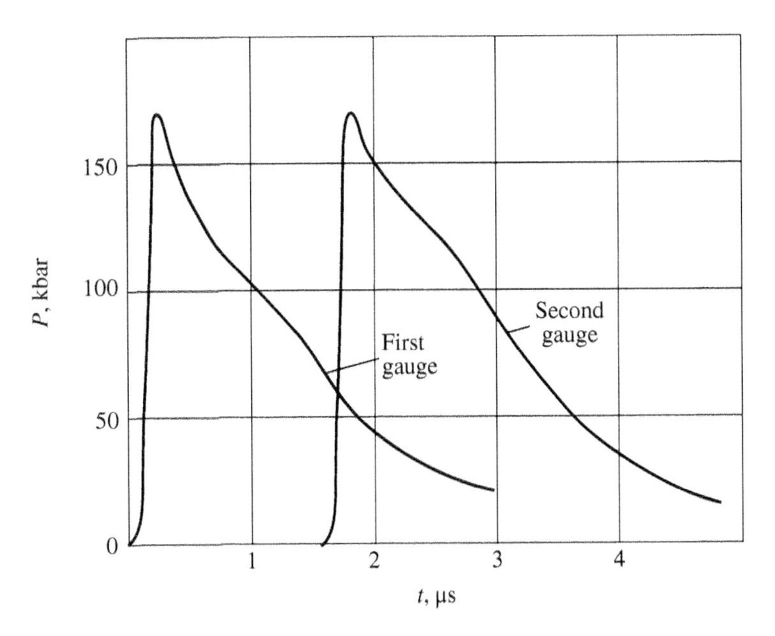

Figure 3.5 Pressure profiles recorded with manganin gauges. Cast TNT/RDX 36/64 (TG-40) without a gasket.

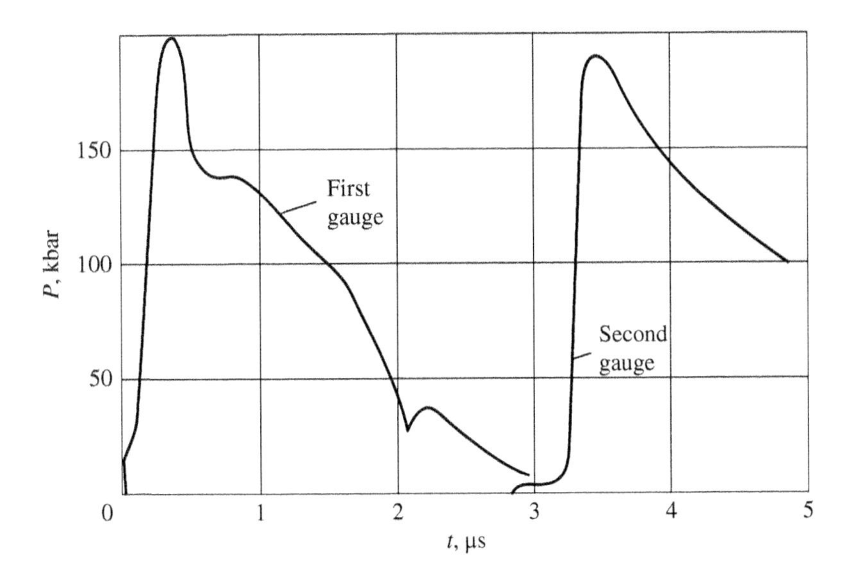

Figure 3.6 Pressure profiles recorded with manganin gauges. Phlegmatized RDX (A-IX-1) with an aluminum gasket of 1.34 mm.

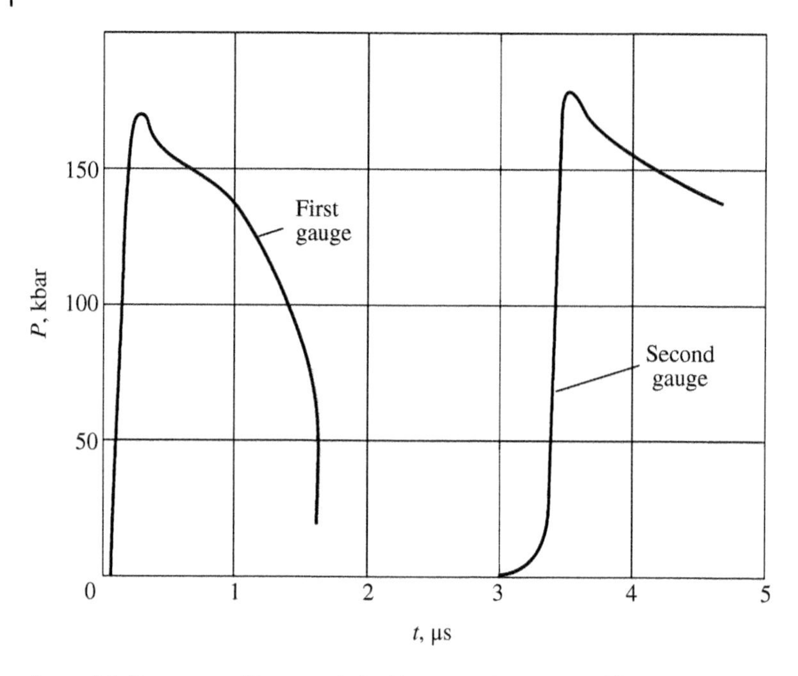

Figure 3.7 Pressure profiles recorded with manganin gauges. Phlegmatized RDX (A-IX-1) without a gasket.

with an aluminum gasket and without a gasket are presented in Figures 3.6 and 3.7, respectively.

The analysis of the measurements shows that the detonation process near the "explosive/ceramic" boundary has a nonstationary behavior, which is manifested in the variability of the pressure values and in the detonation velocity difference in comparison with the stationary detonation parameters.

It should be noted that the measured pressures are lower than the pressure at the Chapman–Jouguet point, and the detonation velocities are higher than the steady-state detonation velocities. An increase in the detonation velocity of RDX is approximately 12%. A noticeable difference in the behavior of the pressed and cast TNT can apparently be explained by differences in their explosive decomposition kinetics.

3.2 Optical Recording in Streak Mode

In this series of experiments, the influence of silicon carbide plates on the shape of the detonation front was investigated. Plates made of self-bonded silicon carbide with a density of 3.10 g/cm^3 and a free silicon content of 11.5% were used. In all cases, the porosity was less than 2% and the free carbon content was 3%. The estimated value of the Hugoniot elastic limit (HEL) was 10 GPa. Composition TG-40 with a density of 1.65 g/cm^3 and phlegmatized RDX with a density of 1.62 g/cm^3 were used.

Initiation of all charges was carried out by PWGs, which had a shape of conical funnels made of 0.3 mm copper foil. They were accelerated by a 2 mm layer of elastic EVV-34 explosive based on pentaerythritol tetranitrate (PETN). The shape of the detonation front was determined with the help of the SFR-2M rotating mirror camera in streak mode. The process of the detonation front exiting to the rear face of the explosive charge was recorded. The recording speed on the film in all cases was 3.75 mm/μs.

Figures 3.8–3.11 show the schemes and results of experiments for composition TG-40.

Figure 3.8 shows the streak camera image of the detonation front in an assembly consisting of a TG-40 charge with 4 mm brass plate on top. The shape of the detonation front near the free surface of the charge does not have any specific peculiarities. However, in the region adjacent to the brass plate, there is a local perturbation. In this and several other experiments, the reason for such a perturbation in the shape of the detonation front was the presence of small gaps (0.1–0.2 mm) between the charge and the plate. These results were used in the analysis of subsequent

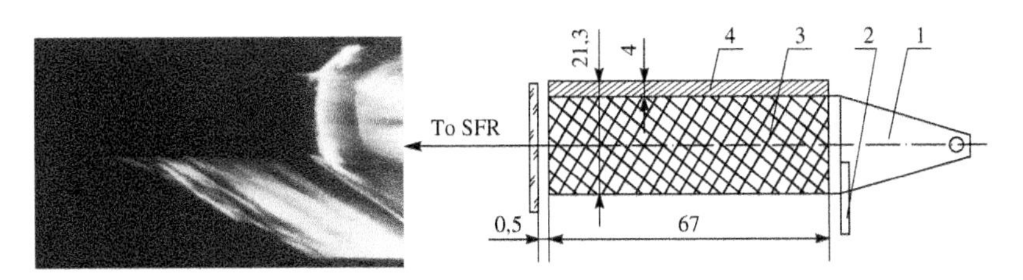

Figure 3.8 Streak camera photograph of the detonation front shape: 1, PWG; 2, detonation start marker; 3, HE TG-40; and 4, brass plate.

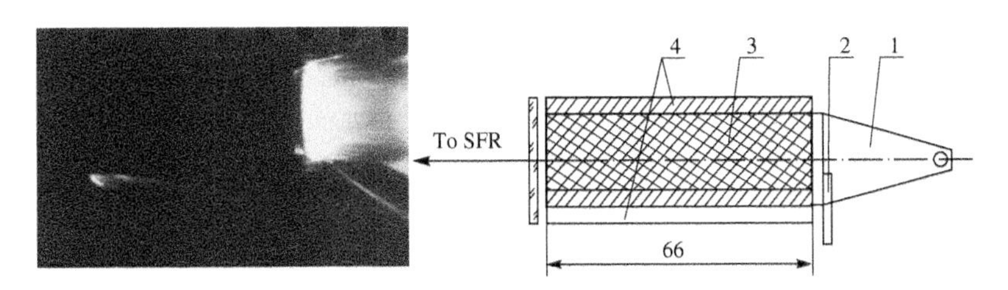

Figure 3.9 Streak camera photograph of the detonation front shape: 1, PWG; 2, detonation start marker; 3, HE TG-40; and 4, brass plates.

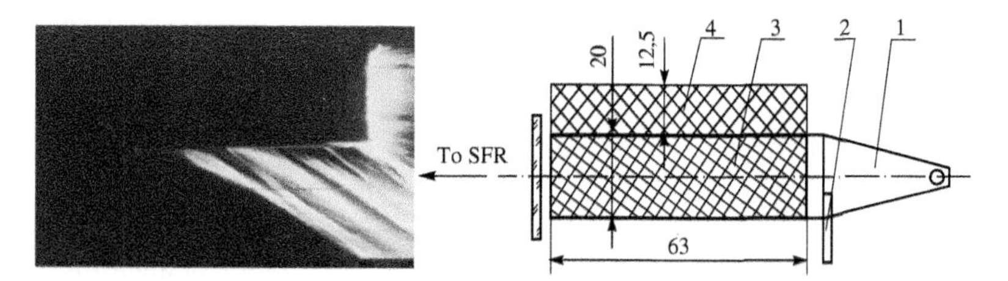

Figure 3.10 Streak camera photograph of the detonation front shape: 1, PWG; 2, detonation start marker; 3, HE TG-40; and 4, silicon carbide plate.

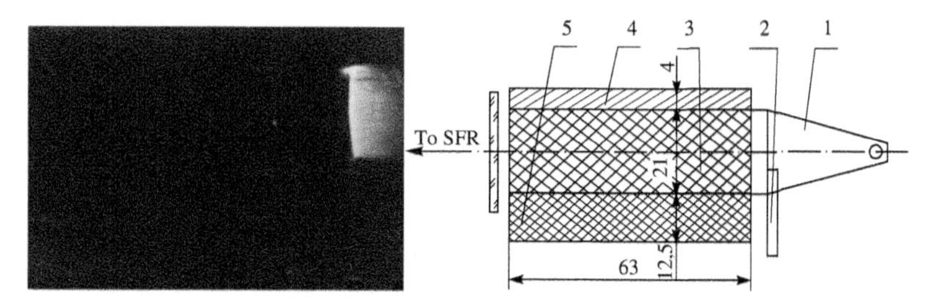

Figure 3.11 Streak camera photograph of the detonation front shape: 1, PWG; 2, detonation start marker; 3, HE TG-40; 4, brass plate; and 5, silicon carbide plate.

experiments. It was concluded that the attribute typical of charges with gaps is a violation of the front symmetry and outrunning glow at the periphery of the charge, which appears due to the channel effect.

Figure 3.9 shows the streak camera photograph of the detonation front in a charge placed between two 4 mm brass plates. The shape of the detonation front, in this case, does not have any significant features.

Streak camera photograph of a charge with a silicon carbide plate on top of it is shown in Figure 3.10. Here, a local delay of the detonation front at the boundary with the ceramic plate is observed. No signs of the channel effect were observed in this case.

Figure 3.11 shows the streak camera photograph for a charge of explosive placed between two plates – a brass plate on top and a silicon carbide plate at the bottom. A noticeable delay of the detonation front propagating to half the height of the charge is observed at the boundary with the silicon carbide plate. At the boundary with the brass plate, a clear manifestation of the channel effect may be noticed.

Figure 3.12 shows a streak camera photograph for a charge placed between two silicon carbide plates. On the boundaries with both the upper and the lower plates there is an evidence of the channel effect. The detonation front itself has a symmetrical shape convex toward the direction of the detonation propagation.

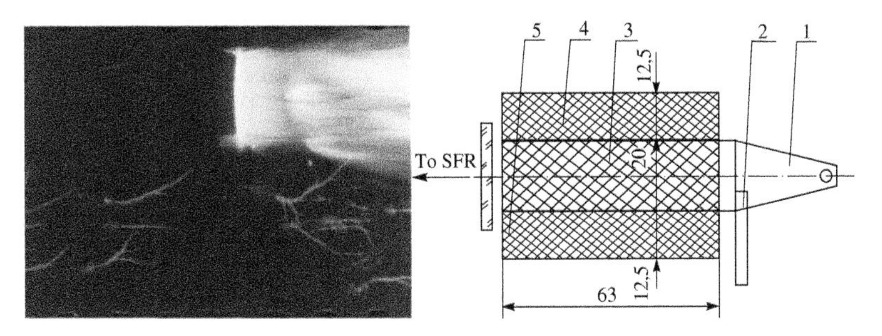

Figure 3.12 Streak camera photograph of the detonation front shape: 1, PWG; 2, detonation start marker; 3, HE TG-40; and 4, silicon carbide plate.

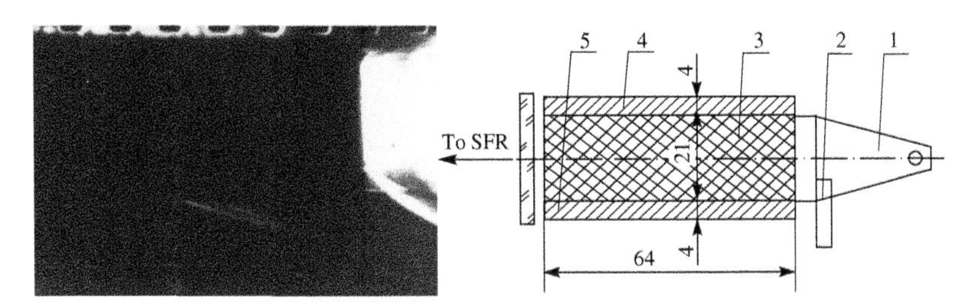

Figure 3.13 Streak camera photograph of the detonation front shape: 1, PWG; 2, detonation start marker; 3, phlegmatized RDX; and 4, 5, brass plates.

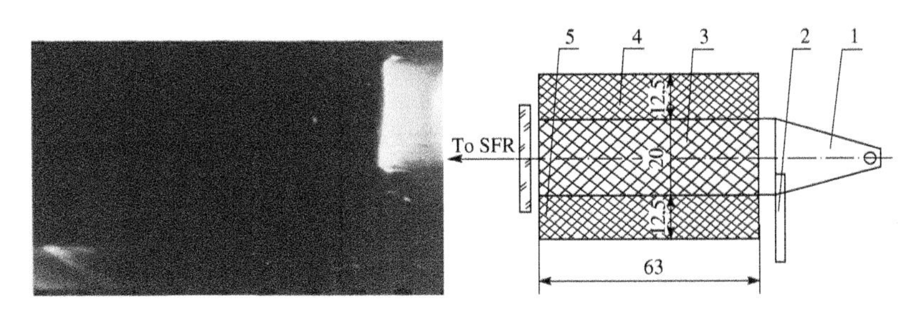

Figure 3.14 Streak camera photograph of the detonation front shape: 1, PWG; 2, detonation start marker; 3, phlegmatized RDX; and 4, 5, silicon carbide plates.

The results previously mentioned convincingly show that for a given explosive, compression due to the outrunning shock wave propagating through the ceramics leads to a marked decrease in the detonation velocity. The effect of compression extends almost to the middle of the charge.

Figures 3.13 and 3.14 present the results of similar experiments for phlegmatized RDX. Figure 3.13 shows a streak camera image of the detonation front in a charge having brass plates on top and bottom. In this case, no distortions of the front can be observed.

The streak camera image of the detonation front in a charge placed between two silicon carbide plates is shown in Figure 3.14. The front has a concave shape relative to the direction of detonation propagation. The results indicate that the compression of a phlegmatized RDX charge by an outrunning shock wave propagating over the ceramic leads to a certain increase in the detonation velocity.

3.3 Modeling of Detonation in High Explosive Charges Contacting with Ceramic Plates

To understand the details of experimental observations, a numerical simulation was performed. Initially, calculations were performed using the software package "POTOK" [12] in a two-dimensional (2D) plane formulation without consideration of the possible material failure. Wilkins model of an elastic plastic medium [13] was used to describe the behavior of the ceramic plate material. In this model, the spherical part of the stress tensor was calculated from the shock adiabat, and the deviatoric component was calculated according to Hooke's law and the law of plastic flow associated with the von Mises yield criterion. The density of the ceramic material was $\rho_M = 3.1\,\mathrm{g/cm^3}$, shear modulus $G = 170\,\mathrm{GPa}$, and HEL = 8 GPa.

The following form of the shock adiabat equation was used in calculations:

$$D = 8.0 + 0.95u,$$

where D is the shock wave velocity and u is the particle velocity behind the detonation front. To model the detonation process in the explosive, the equations of gas dynamics were additionally complemented by the equation of its decomposition kinetics,

$$\frac{\partial \alpha}{\partial t} = K(\alpha + \alpha_0)(p - p*),$$

where α is the degree of explosive decomposition, K and α_0 are parameters of the kinetic equation, p is the pressure, and p^* is the pressure at which the decomposition of explosives begins. Calculation of the parameters of the mixture was performed based on the assumption of additivity of the mass and internal energy within the framework of the single-velocity adiabatic model. In all cases, the spatial resolution of the computational mesh was 1 cell/mm (Figure 3.15). The ceramic plate had a thickness of 10 mm. The left and right boundaries of the calculation area were open, the upper boundary was absent, and the lower boundary

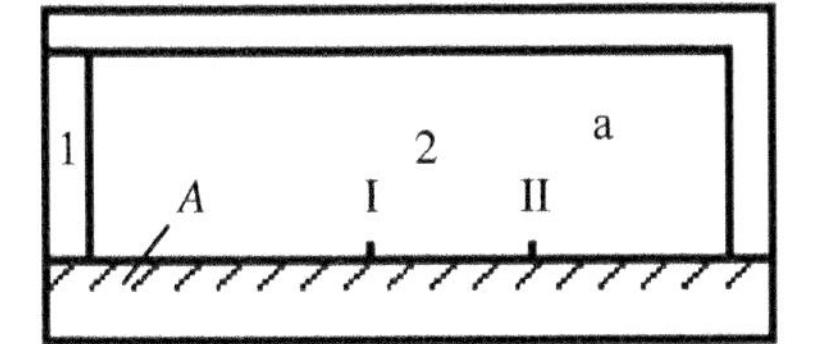

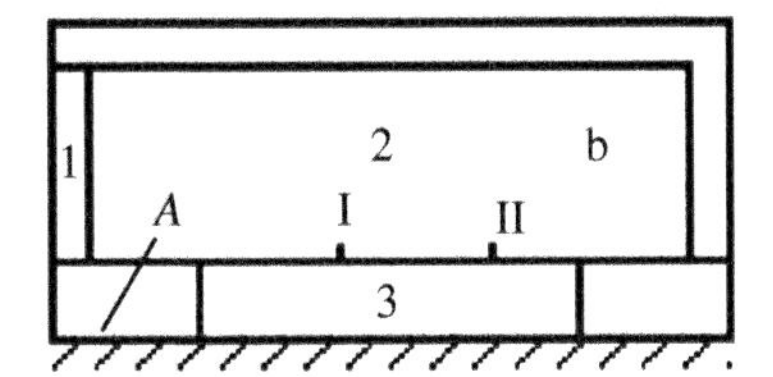

Figure 3.15 Initial configurations of the calculation areas: a, K1 and K3; b, K2 and K4: 1, aluminum impactor (with velocity of 2 km/s); 2, HE; and 3, ceramic plate. A is a rigid wall; I and II are Lagrangian gauges. (*Source:* From Balaganskii et al. [10]. Reprinted with permission of Springer Nature.)

Table 3.2 Parameter values for detonation calculation.

Problem	Ceramic plate	a_0	p^*, GPa
K1	No	0.07	1.3
K2	Yes	0.07	1.3
K3	No	0.10	0.4
K4	Yes	0.10	0.4

Source: From Balaganskii et al. [10]. Reprinted with permission of Springer Nature.
The polytropic index of detonation products is $n = 3.13$; initial density of explosive charge $\rho_0 = 1.6\,\text{g/cm}^3$; sound velocity in explosive $c_{\text{HE}} = 2.5\,\text{km/s}$; and $K = 1.01\ (\text{GPa}\cdot\mu\text{s})^{-1}$.

was closed (i.e. it was considered a rigid wall). The boundaries of open type limited the movement of the adaptive Eulerian mesh and simultaneously allowed the particles of matter to flow inward and leave the simulation area and did not represent obstacles to the propagation of shock waves and other perturbations. The absence of an upper boundary allowed the Eulerian mesh to move along with the deforming body. The boundary conditions on the contact interfaces were defined in such a way that allowed the particles to slip relative to each other. Table 3.2 summarizes the initial data used to set up the simulation.

The results of the modeling are shown in Figures 3.16 and 3.17. The main calculated parameters of the detonation process for all cases are given in Table 3.3.

Analysis of the results shows that the presence of a high-modulus ceramic plate leads to a nonstationarity of detonation process near the interface area. In the calculated region, the detonation velocity increases, and the pressure and particle velocity decrease. The effect is more pronounced for more sensitive explosives, which is in good agreement with the experimental data. The perturbations of the parameters at the interface catch up with the detonation front and change its shape, as shown in Figure 3.16.

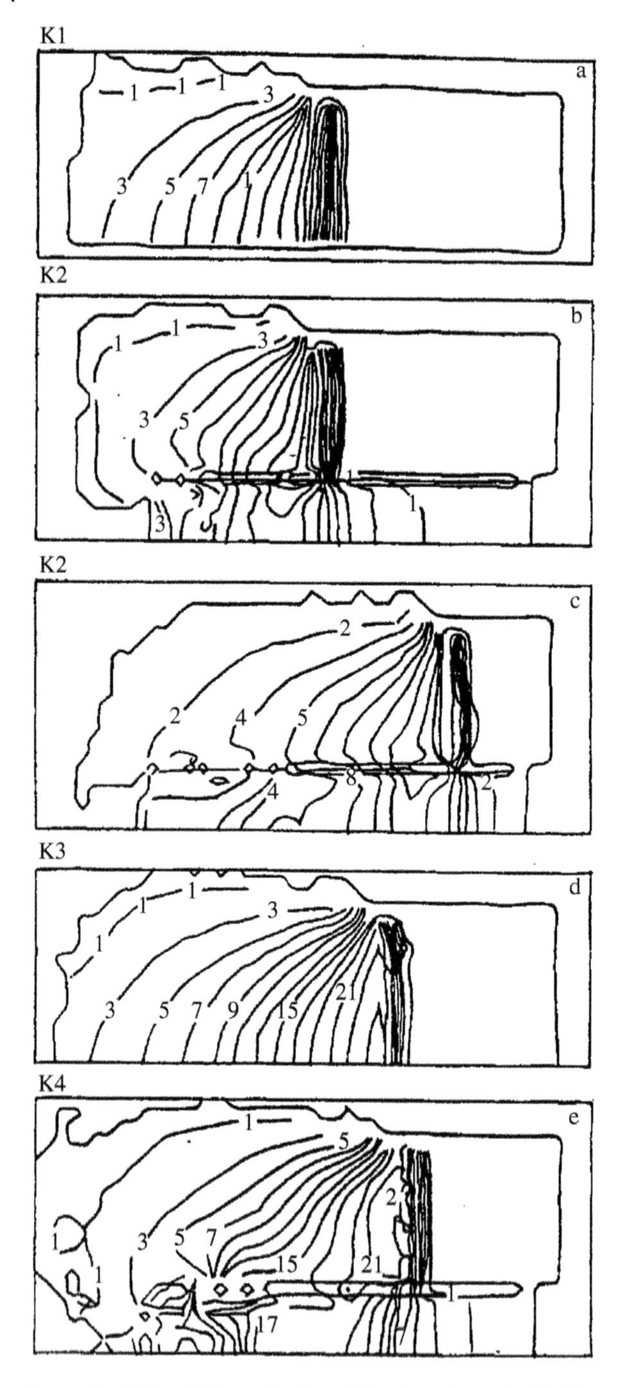

Figure 3.16 Flow fields and isobars (GPa) for variants K1–K4. Time *t*: *a*, *b*, *d*, *e*, 6 μs; *c*, 9 μs. (*Source:* From Balaganskii et al. [10]. Reprinted with permission of Springer Nature.)

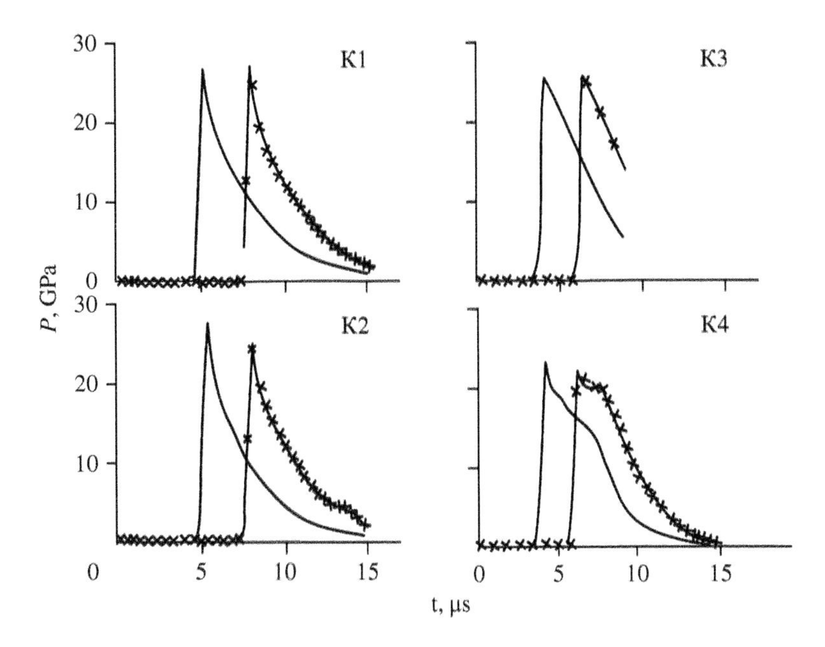

Figure 3.17 Pressure profiles in Lagrangian gauges I and II. (*Source:* From Balaganskii et al. [10]. Reprinted with permission of Springer Nature.)

Table 3.3 Flow parameters at the detonation wave front.

Problem	*D*, km/s	*u*, km/s	*P*, GPa
K1	7.3	2.13	26.6
K2	7.4	1.86	24.0
K3	9.0	1.70	25.0
K4	10.0	1.13	23.0

Source: From Balaganskii et al. [10]. Reprinted with permission of Springer Nature.

A more detailed representation of the process was produced by a simulation using the AUTODYN software in a 2D formulation on the Eulerian mesh. Detonation of a TG-40 high explosive (HE) charge placed between plates of copper and silicon carbide was thoroughly investigated. The scheme of the computational area at the initial moment of time is shown in Figure 3.18. All the initial data and the equations of state were similar to those previously used in AUTODYN simulations. The spatial resolution was 10 cells/mm.

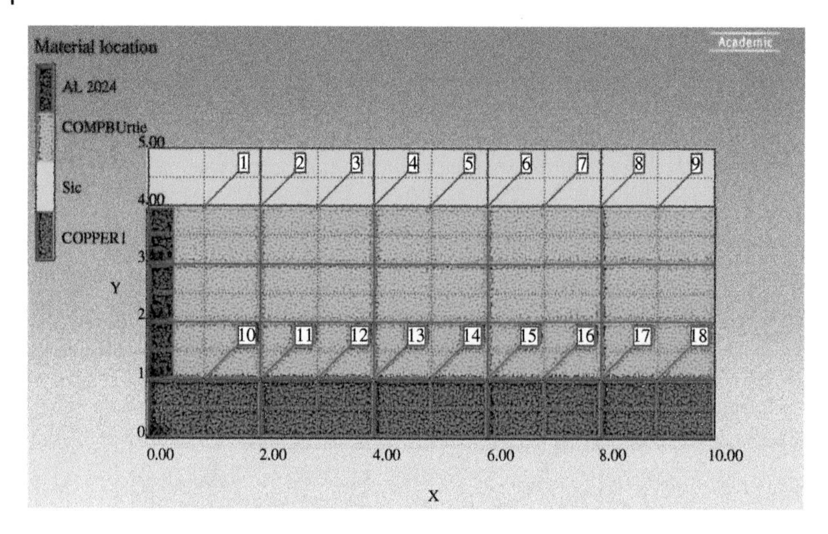

Figure 3.18 Scheme of the computational area.

On the boundaries between the charge and inert walls with an interval of 1 cm, Lagrangian gauges were placed, in which the change in the parameters as a function of time was recorded. The gauges numbered from 1 to 9 were on the boundary with the ceramic plate, while those from 10 to 18 were placed at the boundary with a copper plate. Initiation of the setup was carried out by an impact of the aluminum plate on the left end of the charge with a velocity of 2.5 km/s.

The calculated flow fields and the velocity vectors of the particles at different times in increments of 2 μs are shown in Figure 3.19. The presented data clearly demonstrate the peculiarities of the development of the detonation process near the bounding plates of different materials.

A nearly stationary oblique shock wave propagates in a copper plate, which is reflected as a rarefaction wave transmitted to the products of the detonation. The shock wave in the ceramic plate is significantly ahead of the detonation front and compresses the explosive in a region adjacent to the plate.

Figure 3.20 shows a comparison of the detonation transformation durations at the gauge locations. The vertical axis of the diagram shows the number of the gauge, while the horizontal axis corresponds to the time (μs) from the moment of detonation initiation. Dark rectangles represent the time of complete conversion of explosives into detonation products.

The duration of the transformation near the boundary with ceramics increases in comparison with that near the copper plate; however, the reaction near ceramic plate does not start earlier.

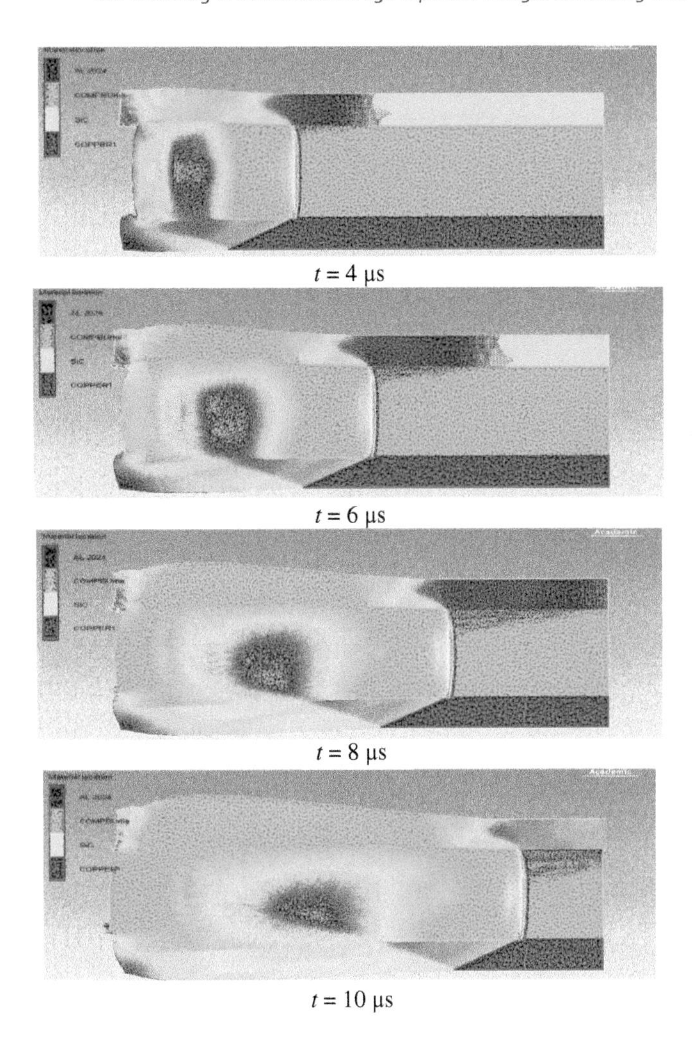

Figure 3.19 Calculated flow fields and the velocity vectors of the particles at different times.

Figure 3.21 illustrates the change in detonation parameters as the process propagates through the charge.

The paper [14] aims at analyzing the refraction phenomena of detonation waves in condensed HEs in theoretical and numerical ways. According to these results for PBX9502 explosive interacting with beryllium interface, four types of refraction of the detonation wave at the high sound-speed material interface can be identified: a regular refraction with a reflecting shock wave, an irregular refraction with a bound precursor

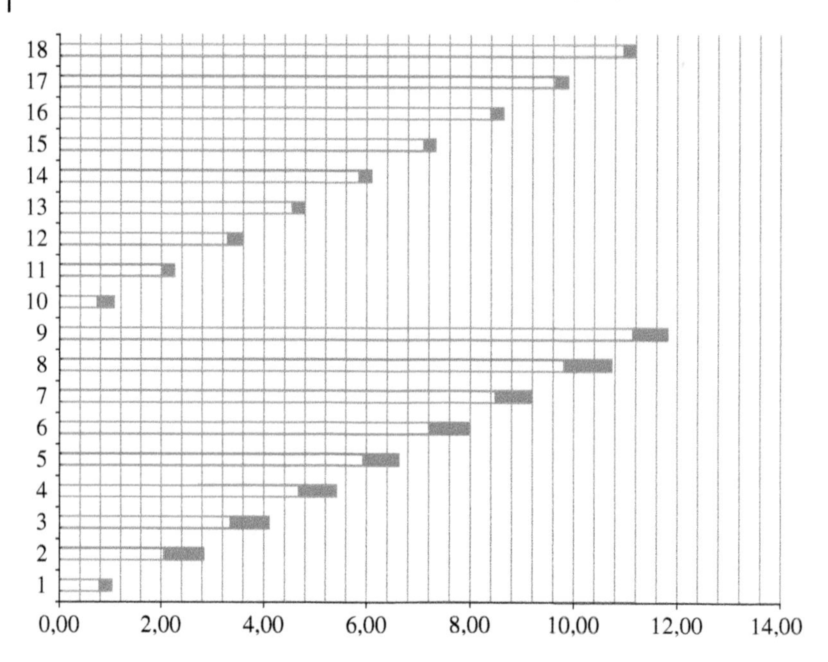

Figure 3.20 Comparison of the detonation transformation durations at the locations of the gauges.

wave, an irregular refraction with a twin Mach reflection, and an irregular refraction with a λ-wave structure (see Figure 3.22). In the first type, the front of the leading shock wave is straight, the flows in the detonation reactive zone and beryllium are both supersonic, and the reflecting shock wave appears behind the leading shock wave and a refracting shock wave appears within beryllium. In the second type, the front of the leading shock wave is also straight, and the flow in the detonation reactive zone is supersonic but the one in beryllium is subsonic, so a reflecting shock wave appears behind the leading shock wave and a refracting shock wave appears within beryllium, too; moreover, the refracting shock wave is almost perpendicular to the material interface, that is a bound precursor wave.

For the third type, the front of the leading shock wave becomes a forward curve, and the flows in the detonation reactive zone and beryllium are both subsonic, i.e. a Mach item is produced at some distance above the material interface where there are two Mach reflection structures on the top and the bottom of the Mach item, respectively. Obviously, the bottom Mach reflection is a free precursor wave from the refracting shock wave within beryllium. For the fourth type, the forward curve range of the front of the leading shock wave becomes very broad, and

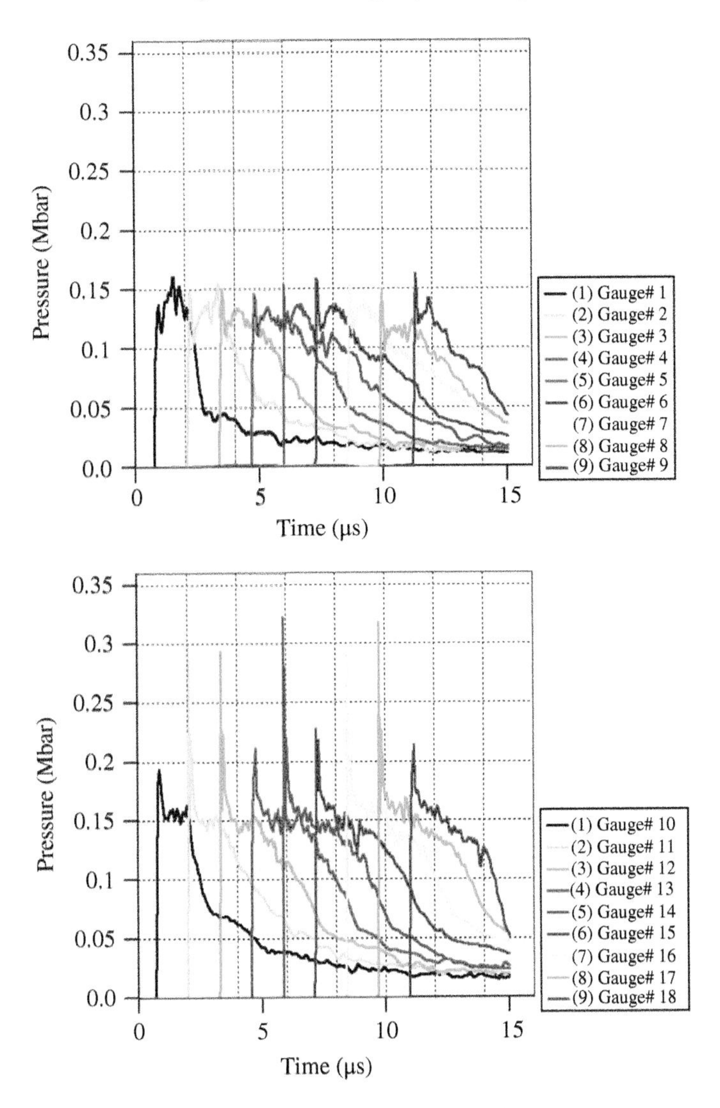

Figure 3.21 Change in detonation parameters as the process propagates through the charge (top: ceramic and bottom: copper plate).

accordingly the range of subsonic flows in the detonation reactive zone becomes very wide. This makes the top Mach reflection disappear but the bottom one still exists, so the whole structure of the reflection wave looks like the Greek letter λ; meanwhile, the flows within beryllium may be all in a subsonic state.

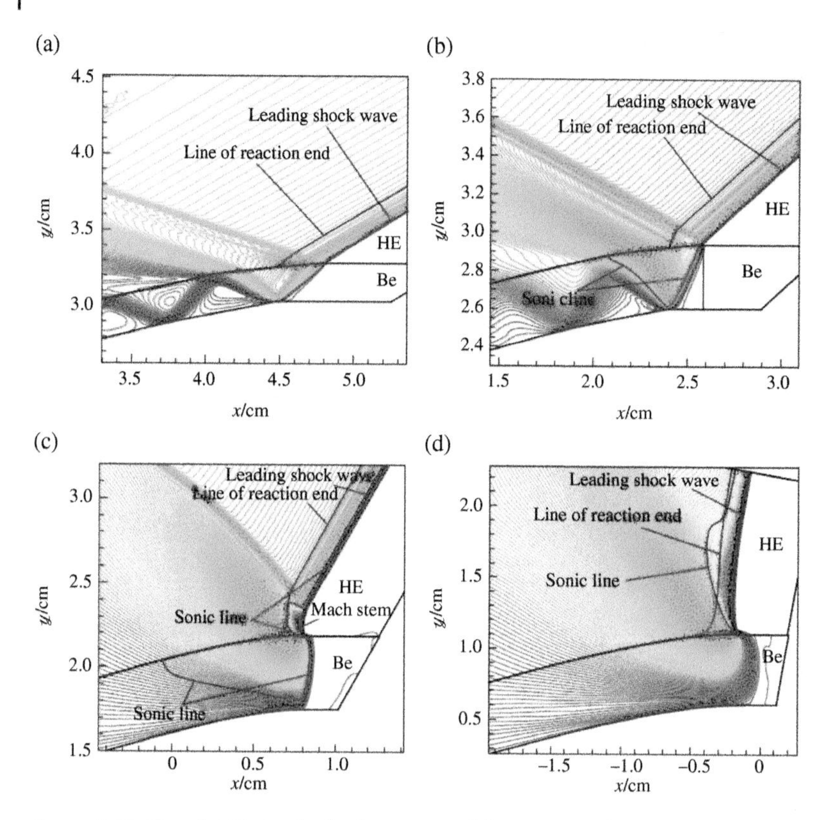

Figure 3.22 Flow field behavior for four types of detonation refraction.

3.4 Summary

The results of experimental studies of the detonation process propagation at the interface between explosive charges and inert plates have shown that in the case when the sound velocity in an inert material is less than the detonation velocity, the plate does not affect the stationary behavior of the detonation transformation. The data obtained with the help of manganin gauges indicate that the detonation process near the explosive/ceramics boundary has a pronounced nonstationary behavior, which manifests itself in the variability of the pressure values and in the difference of detonation velocity from that typical for the stationary detonation. For the phlegmatized RDX, the measured pressures are lower than the pressure at the Chapman–Jouguet point, and the detonation velocities are higher by about 12% than the stationary detonation velocities. A noticeable difference in the behavior of pressed and cast TNT can apparently be explained by the differences in the kinetics of explosive decomposition. The results for TG-40 show that for a given explosive, the compression of a leading shock wave propagating through the ceramics leads to an appreciable decrease in the detonation velocity. The effect of compression extends almost to the half the height of the charge.

The numerical analysis confirmed that the presence of a high-modulus ceramic plate leads to the nonstationarity of the detonation process at the interface. This effect is more pronounced for more sensitive explosives, which agrees with the experimental data. In this case, perturbations arising at the interface affect the front of the detonation wave and change its shape.

The duration of the detonation transformation near the ceramic plate increases in comparison with the duration near the copper plate, but the reaction does not start earlier.

The study of the theoretical and numerical results of the refraction of detonation waves in PBX9502 explosive interacting with beryllium interfaces carried out in China has identified four types of refraction of the detonation wave at the high sound-speed material interface: a regular refraction with a reflecting shock wave, an irregular refraction with a bound precursor wave, an irregular refraction with a twin Mach reflection, and an irregular refraction with λ-wave structure.

References

1 Stanyukovich, K.P. (1955). *Unsteady Motion of a Continuous Medium.* Moscow: Gostekhizdat [in Russian].
2 Ovsyannikov, L.V. (1981). *Lectures on the Fundamentals of Gas Dynamics.* Moscow: Nauka [in Russian].
3 Merzhievskii, L.A., Fadeenko, Y.I., Filimonov, V.A. et al. (1976). Acceleration detonation propagation in charges with a litium-filled hollow. *Combustion, Explosion and Shock Waves* 12 (2): 205–211.
4 Merzievski, L.A., Fadeenko, Y.I., and Chistjakov, V.P. (1977). Detonation of cylindrical charge with lithium-filled cavity. *Acta Astronautica* 4: 459–496.
5 Zababakhin, E.I. (1997). *Some Problems of the Gas Dynamics of the Explosion.* Snezhinsk: RFNC [in Russian].
6 Eden, G. and Belcher, R.A. (1989). The effects of inert walls on the velocity of detonation in EDC 35, an insensitive high explosive. *Proceedings of the 9th International Symposium on Detonation*, Portland, USA (28 August–1 September 1989). Arlington: Office of Naval Research.
7 Aveille, J., Carion, N., Vacellier, J. et al. (1989). Experimental and numerical study of oblique interaction of detonation waves with explosive/solid material. *Proceedings of the 9th International Symposium on Detonation*, Portland, USA (28 August–1 September 1989). Arlington: Office of Naval Research.

8 Balagansky, I.A., Berdnik, V.P., Kulikova, I.V. et al. (1991). Features of detonation processes in HE charges that contact with high modulus ceramics. *Annotations of the Reports of the 7th All-Union Congress on Theoretical and Applied Mechanics*. Moscow, USSR (15–21 August 1991). Moscow: Moscow State University. [in Russian].

9 Balagansky, I.A., Razorenov, S.V. and Utkin, A.V. (1993). Detonation parameters of condensed high explosive charges with long ceramic elements. *Proceedings of the 10th International Detonation Symposium*, Boston, USA (12–16 July 1993). Arlington: Office of Naval Research.

10 Balaganskii, I.A., Agureikin, V.A., Razorenov, S.V. et al. (1994). Effect of an inert high-modulus ceramic wall on detonation propagation in solid explosive charges. *Combustion, Explosion and Shock Waves* 30: 674–681.

11 Balagansky, I.A., Balagansky, A.I., Kobilkin, I.F. et al. (2005). Influence of high explosive charge shell on detonation front shape. *Proceedings of International Conference 'VII Zababakhin Scientific readings'*, Snezhinsk, Russia (5–9 September 2005). Snezhinsk: RFNC. [in Russian].

12 Agureikin, V.A., Vopilov, A.A., and Kulkov, O.N. (1987). *The software package POTOK-ES. User Manual*. Novosibirsk: Institute of Applied Physics [in Russian].

13 Wilkins, M.L. (1964). Calculation of elastic-plastic flows. In: *Methods in computational physics*, vol. 3 (ed. B. Alder, S. Fernbach and M. Rotenberg), 211–263. New York/London: Academic Press.

14 Ming, Y. and Quan, L. (2016). Refraction of detonation wave at interface between condensed explosives and high sound-speed material. *Acta Physica Sinica* 5 (2): 024702. [in Chinese].

4

Peculiar Properties of the Processes in High Explosive Charges with Cylindrical Shells

In the first chapter, we have already discussed the processes that develop in cylindrical explosive charges. The interest in charges encapsulated in shells of inert materials is associated with several problems related to the physics of explosions. First, the critical diameters of explosives placed in shells are much smaller than for open charges. Several studies of the effect of shells on critical diameters are described in monographs on explosion physics. In such cases, a clear understanding of the interaction of the detonation wave front with the shell is of fundamental importance. Besides, the results of experiments on throwing shells by explosion are actively used to derive and test the equations of state for detonation products (some examples of such studies may be found in [1–3]). From a practical point of view, shell fracturing and fragment spreading processes are investigated in [4]. This chapter deals with the peculiarities of detonation processes in explosive charges placed in shells made of ceramic materials.

4.1 Nonstationary Detonation Processes in High Explosive Charges with Silicon Carbide Shells

As in the case of detonation in charges contacting with ceramic plates, the first series of experiments was conducted with the help of manganin gauges [5–7]. Nitromethane (density = 1.14 g/cm^3), phlegmatized hexogen (A-IX-I) of gravimetric density, cast trinitrotoluene (TNT), and a composition of TNT with RDX (TG-40) having densities equal to 1.60 and 1.67 g/cm^3, respectively, were used as explosives.

The length of the shell l was 60 mm and the outer diameter D and inner diameter d were 74 and 48 mm, respectively (Figure 4.1). Several

Explosion Systems with Inert High-Modulus Components: Increasing the Efficiency of Blast Technologies and Their Applications, First Edition. Igor A. Balagansky, Anatoliy A. Bataev, and Ivan A. Bataev.

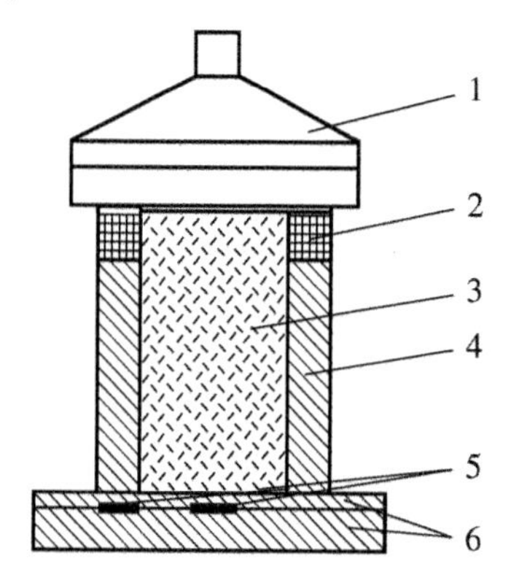

Figure 4.1 The scheme of the experiment: 1, plane wave generator; 2, polystyrene ring; 3, HE; 4, ceramic shell; 5, manganin gauges; and 6, copper screens.

experiments with shells of other sizes were also conducted. The information about these experiments is reflected in the captions of the figures containing the results of corresponding measurements. The inner surface and the front and rear ends of all shells were polished. The gauges were located on the axis of the charge and under the rear end surface of the shell. The size of the gauges was 5 × 5 mm. The fluoroplastic insulation thickness was 80 μm. Detonation waves in explosive charges and shock waves in shells were excited by means of plane wave generators. The application of plane wave generators ensured that the time needed to achieve the stationary detonation mode in the explosive charge did not exceed 1 μs. In experiments with nitromethane, an intermediate charge of phlegmatized pressed RDX (A-IX-1) with a thickness of 40 mm was used between the plane wave generator and the main charge to ensure reliable detonation initiation. To soften the impact of the detonation products on the front end surface of the ceramic shell, a foam ring was used (see Figure 4.1).

The results of the measurements are shown in Figures 4.2–4.14, where the solid lines correspond to records of the gauges and the dashed lines show the curves obtained by a recalculation of the experimental data using the known shock adiabats and considering the pressure values in explosives and in shells, respectively. The profiles of waves in the ceramic shell recorded by manganin gauges are not fully correct since these gauges were subjected to nonuniform loading. The only value that is correctly recorded is the moment of arrival of the elastic wave on the rear end of the shell. The second peak of pressure detected by the gauges on the axis of symmetry is associated with the circulation of waves in copper screens. Thus, it was not taken into account during recalculation of data for explosives. The experimental error in measured pressures was less than 5%.

It should be noted that in the experiment, where water was placed in the shell instead of an explosive charge (Figure 4.2), the registration of the wave by the gauge under the rear end surface of the shell was correct.

The results of the experiments are summarized in Table 4.1, which shows the Chapman–Jouguet parameters for the stationary detonation of explosives (P_{CJ}, D_{CJ}) and the maximum pressures in high explosive

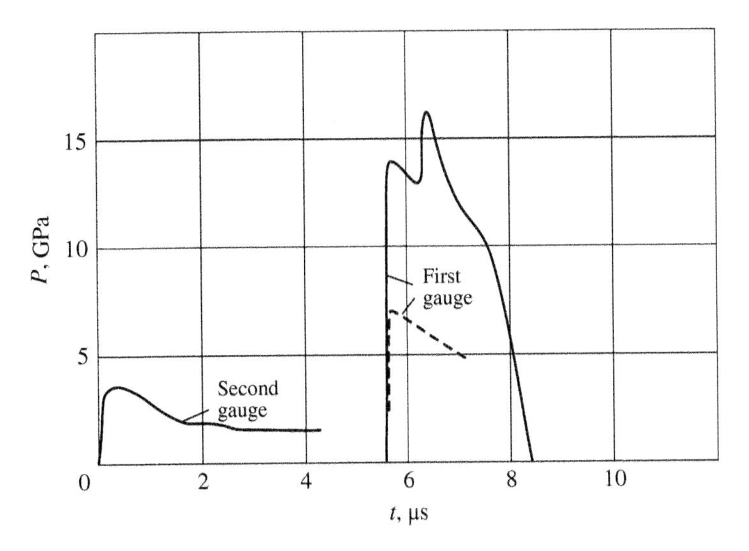

Figure 4.2 Water. Experiment without foam ring: $l = 60$ mm, $D = 74$ mm, and $d = 48$ mm.

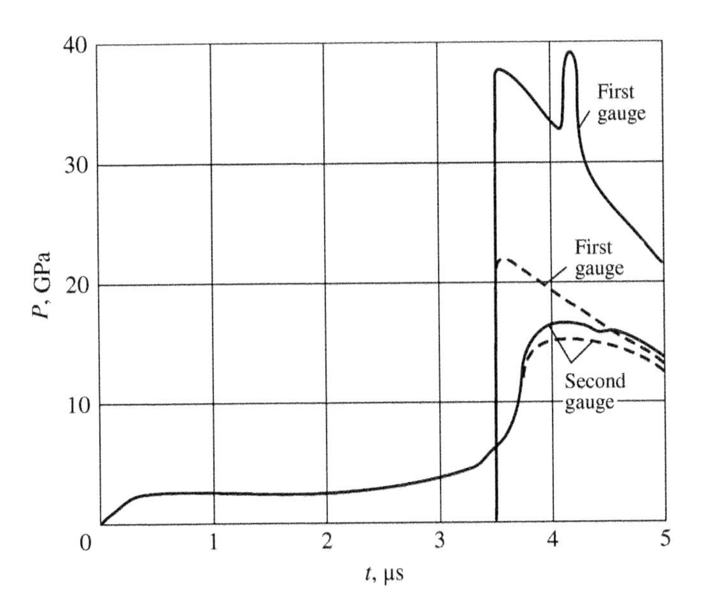

Figure 4.3 Nitromethane. Experiment without foam ring: $l = 60$ mm, $D = 74$ mm, and $d = 48$ mm.

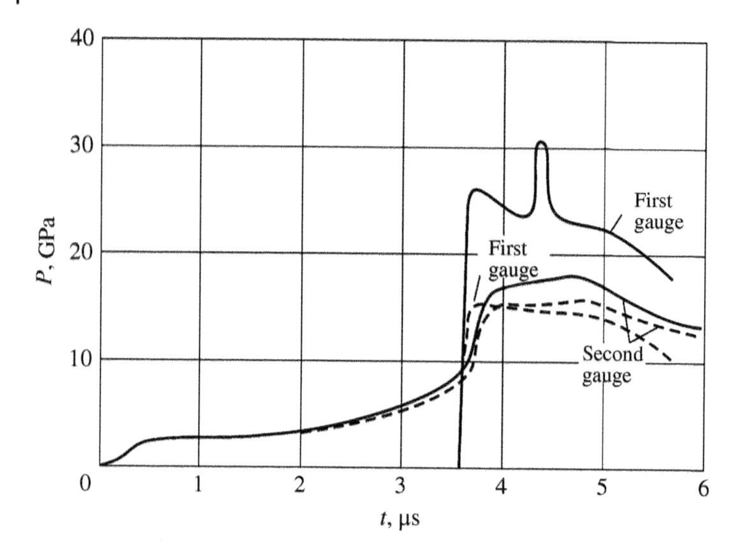

Figure 4.4 Nitromethane. Experiment with foam ring: $l = 60$ mm, $D = 74$ mm, and $d = 48$ mm.

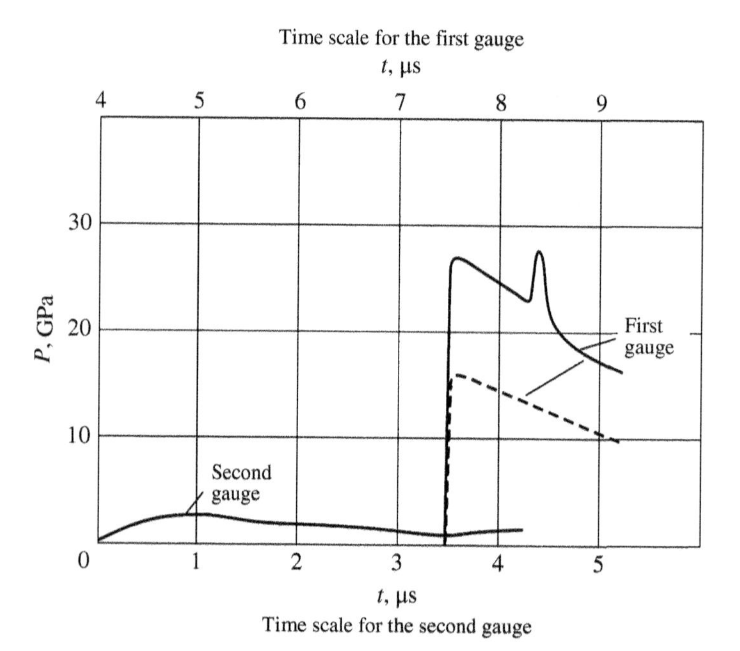

Figure 4.5 Nitromethane. Experiment with foam ring: $l = 120$ mm, $D = 74$ mm, and $d = 48$ mm.

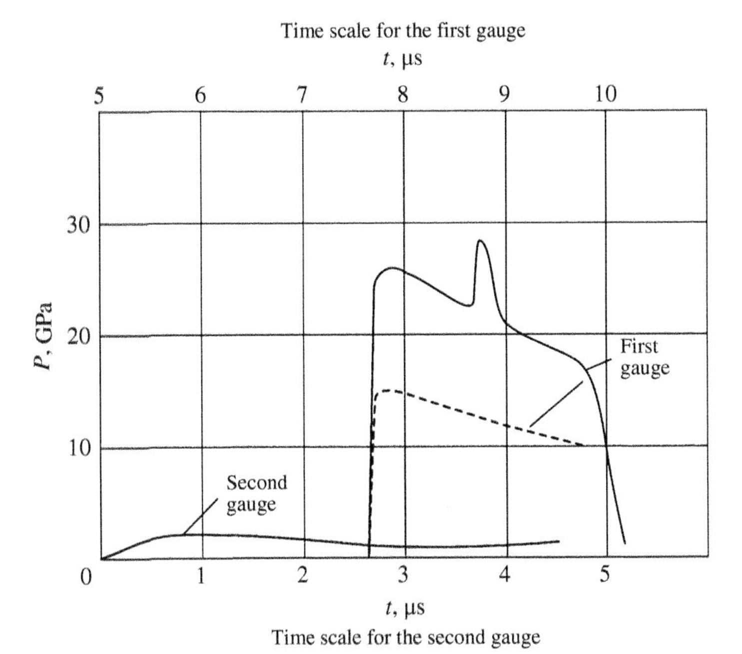

Figure 4.6 Nitromethane. Experiment without foam ring: $l = 120$ mm, $D = 74$ mm, and $d = 48$ mm.

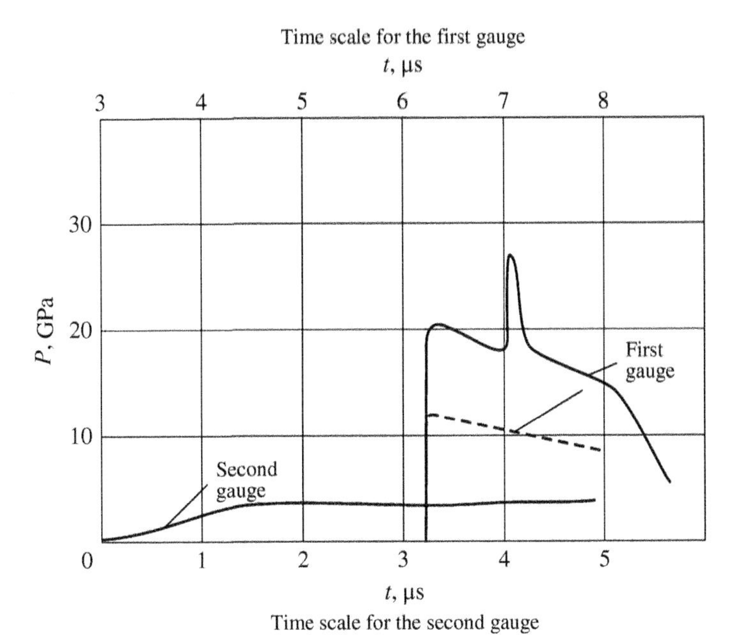

Figure 4.7 Nitromethane. Experiment with foam ring: $l = 98$ mm, $D = 60$ mm, and $d = 40$ mm.

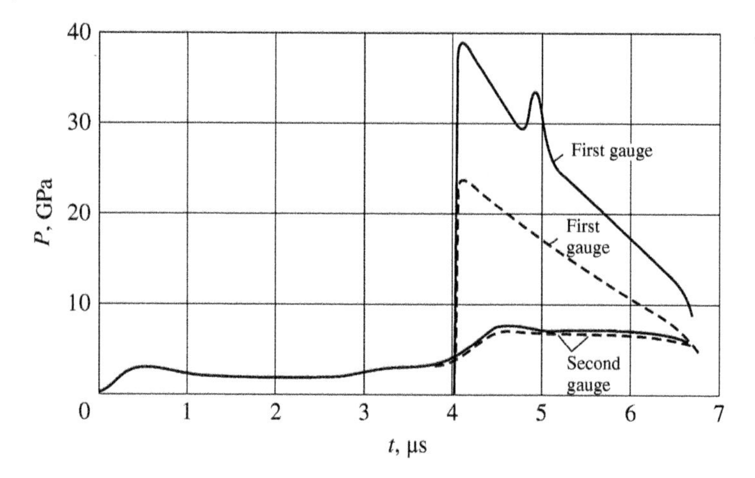

Figure 4.8 Cast TNT. Experiment without foam ring: $l = 60$ mm, $D = 74$ mm, and $d = 48$ mm.

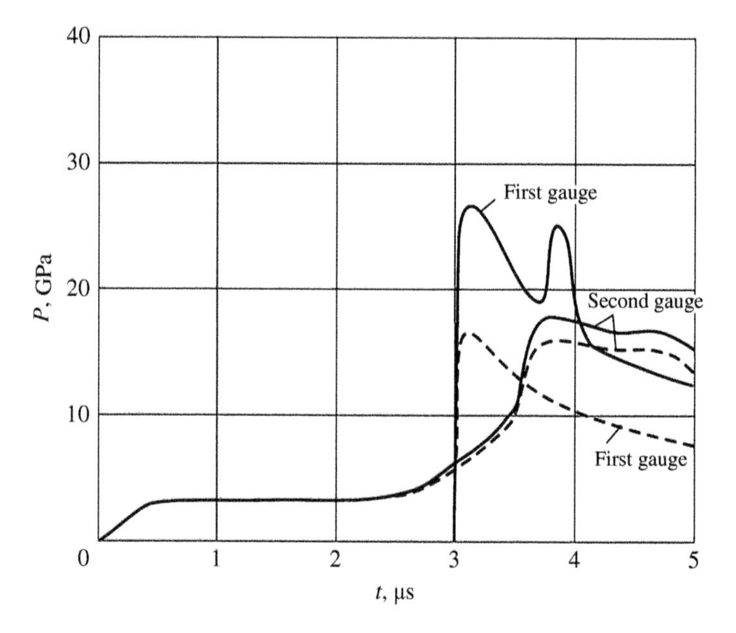

Figure 4.9 Cast TNT. Experiment with foam ring: $l = 60$ mm, $D = 74$ mm, and $d = 48$ mm.

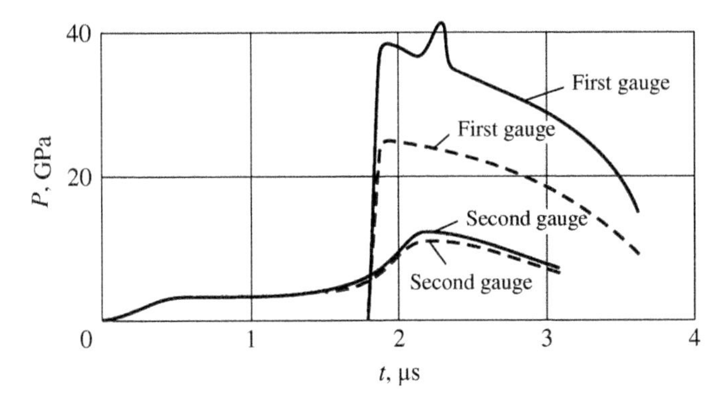

Figure 4.10 Experiment without foam ring: Cast TG-40: $l = 60$ mm, $D = 74$ mm, and $d = 48$ mm.

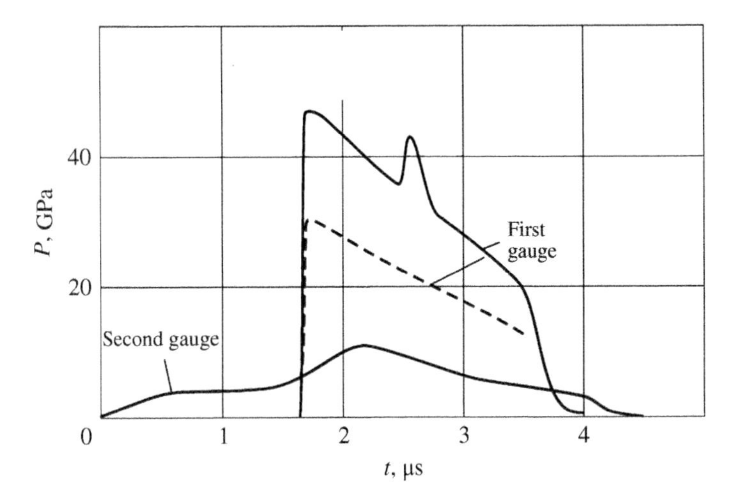

Figure 4.11 Experiment with foam ring: Cast TG-40: $l = 60$ mm, $D = 74$ mm, and $d = 48$ mm.

(P_{HE}) and ceramics (P_C) for shells with dimensions $l = 60$ mm, $D = 74$ mm, $d = 48$ mm.

The analysis of the experimental results shows that the pressures measured by the central gauge depended on the initial conditions. This means that the detonation process did not reach a steady state. For nitromethane, for example, the presence or absence of a foam ring led to significant changes in pressure on the axis of symmetry of the explosive charge. The measured values of the pressure in the detonation wave in most cases differed markedly from the pressures for stationary detonation. In the

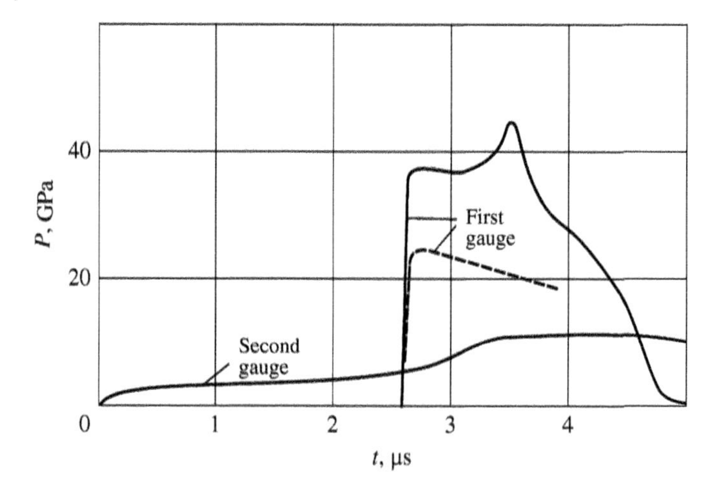

Figure 4.12 Cast TG-40. Experiment with foam ring: $l = 98$ mm, $D = 60$ mm, and $d = 40$ mm.

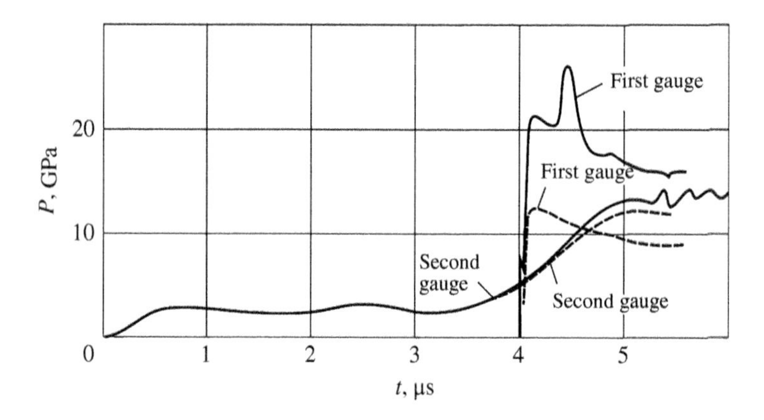

Figure 4.13 Experiment without foam ring: Phlegmatized hexogen (A-IX-1): $l = 60$ mm, $D = 74$ mm, and $d = 48$ mm.

case of nitromethane, a significant pressure increase was observed at the front of the detonation wave, which reached 56% of the Chapman–Jouguet pressure. At the same time, in two experiments, the measured pressures were lower than the Chapman–Jouguet pressures. For instance, a 13% decrease in pressure was observed for composition TG-40. It is important to note that the described effects were obtained on low sensitive explosive compositions. Thus, these effects are not related to the initiation of the explosive by the leading wave in the ceramic shell.

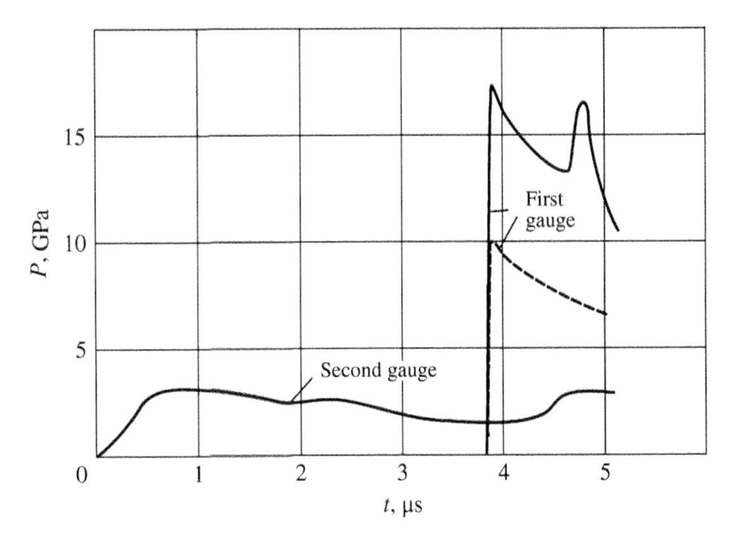

Figure 4.14 Experiment with foam ring: Phlegmatized RDX (A-IX-1): $l = 60$ mm, $D = 74$ mm, and $d = 48$ mm.

Table 4.1 Initial data and test results.

High explosive (HE)	ρ, g/cm^3	Chapman–Jouguet parameters P_{CJ}, GPa	D_{CJ}, km/s	Ring	P_{HE}, GPa	P_C, GPa
Nitromethane	1.14	14.1	6.3	No	22.0	16.5
				Yes	15.5	16.0
Cast trinitrotoluene (TNT)	1.60	19.0	6.9	No	24.0	7.5
				Yes	17.0	15.8
Cast TG-40	1.67	25.0	8.0	No	25.0	11.0
				Yes	30.8	11.2
Phlegmatized hexogen (A-IX-1)	1.10	10.0	6.3	No	12.5	12.0
				Yes	10.0	3.2

In the second series of experiments [8], the detonation processes occurring in the charges surrounded by shells were investigated with the help of optical recording in the streak mode. Hollow cylinders made of self-bonded silicon carbide were used to fabricate the shells. The following substances were used for explosive charges: composition TG-40 with a density of 1.65 g/cm^3, cast TNT with the density of 1.56 g/cm^3, and a cast composition TGA-40/40/20 with addition of aluminum powder. Initiation of all charges was performed by plane wave generators, which were designed as conical funnels made of 0.3 mm copper foil accelerated by a 2 mm layer of the EVV-34 elastic high explosive (HE). The shape of the detonation front was measured using the SFR-2M camera in the streak mode. The output of the detonation front to the end of the explosive charge was recorded. The recording speed on the film in all cases was 3.75 mm/μs.

To compare and emphasize the effects associated with the influence of the ceramic shells, similar experiments were conducted with charges without shells and with brass shells.

Figures 4.15–4.18 show the schemes and experimental results for the composition TG-40 with a density of 1.65 g/cm^3. Figure 4.15 shows a streak camera photograph of the detonation front for a TG-40 charge in a shell of silicon carbide with the point initiation. In this experiment, an intermediate layer of EVV-34 was placed between the detonator and the main charge for reliable initiation.

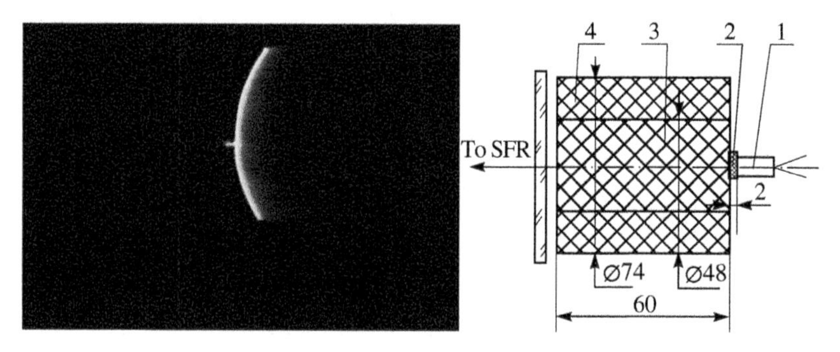

Figure 4.15 Streak camera photograph of the detonation front shape: 1, detonator; 2, EVV-34; 3, TG-40; and 4, silicon carbide shell.

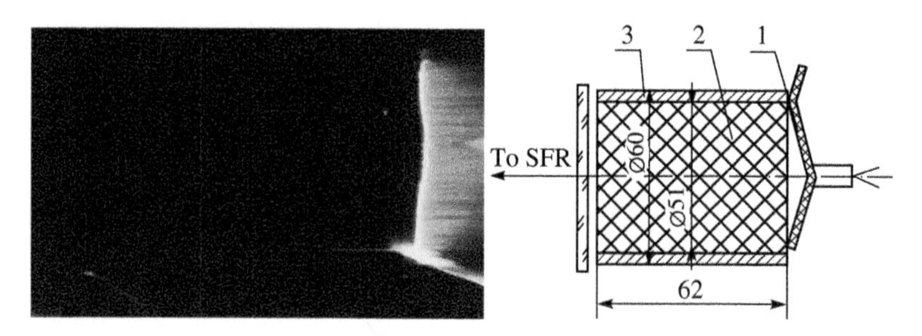

Figure 4.16 Streak camera photograph of the detonation front shape: 1, plane wave generator; 2, TG-40; and 3, brass shell.

It may be noticed that there is a perturbation of the detonation front on the axis of symmetry, which is possibly related to the collision of transverse waves. Figure 4.16 is a streak camera photograph for a TG-40 charge in a brass shell initiated by a plane wave generator. In the lower part of the photograph, there is a noticeable distortion of the detonation front shape (outrunning glow at the boundary with the shell), which is related to the channel effect, which apparently appeared because of the gap between the explosive and the shell.

For the sake of comparison, Figure 4.17 shows the photograph of the detonation front shape in a charge placed in a silicon carbide shell. One may observe distortions on the detonation front that manifest themselves in the form of transverse waves.

Figure 4.18 shows the detonation front shape in the TG-40 charge placed in a shell two times longer than the one shown in Figure 4.17. In this case, the shape of the front noticeably differs from the one shown in the previous figure.

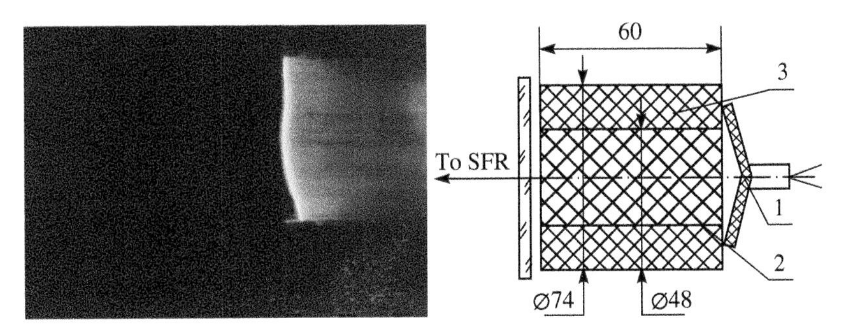

Figure 4.17 Streak camera photograph of the detonation front shape: 1, plane wave generator; 2, TG-40; and 3, silicon carbide shell.

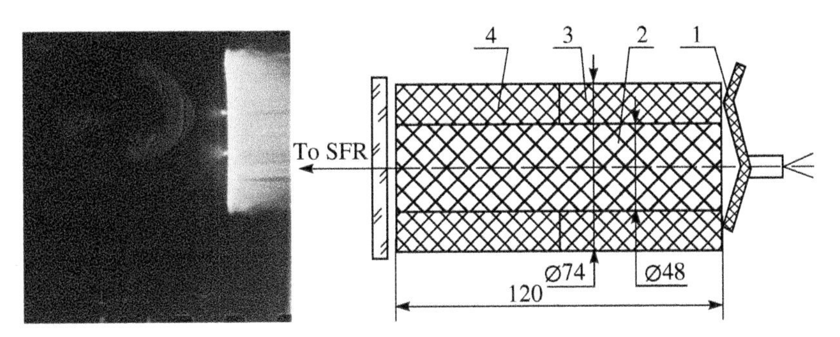

Figure 4.18 Streak camera photograph of the detonation front shape: 1, plane wave generator; 2, TG-40; and 3, silicon carbide shell.

Figures 4.19–4.22 show the results of experiments with charges of cast TNT with the density of 1.56 g/cm^3. Figure 4.19 shows a photograph for the charge of cast TNT without a shell. An additional charge of EVV-34 was placed between the plane wave generator and the main charge to ensure reliable initiation. One may observe a practically flat front without any peculiarities.

Figure 4.20 shows the shape of the detonation front in the charge of cast TNT in the brass shell. One may notice the perturbations on the detonation front, symmetrically propagating from the periphery to the axis of the explosive charge. Most likely, the source of these disturbances is a well-pronounced channel effect.

Figure 4.21 shows the streak camera photograph of the detonation front in the charge of cast TNT placed in a silicon carbide shell. The front of the wave is strongly distorted, but it is not possible to draw an unambiguous conclusion about the causes of the distortion (the channel effect or the influence of the outrunning wave in the ceramic shell).

The streak camera photograph of the detonation front in the TGA-40/40/20 (TNT-RDX-aluminum powder) charge is shown in

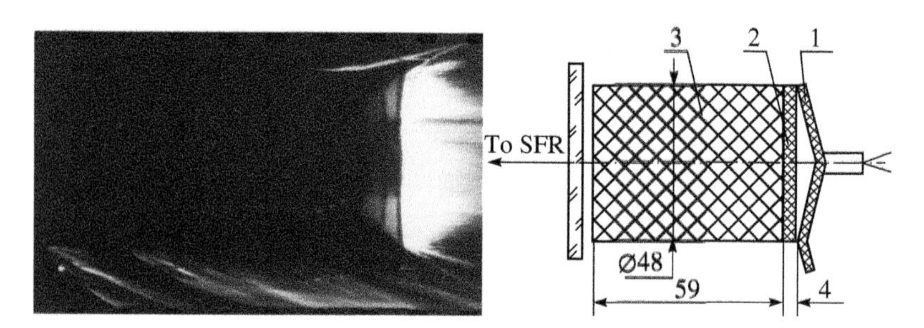

Figure 4.19 Streak camera photograph of the detonation front shape: 1, plane wave generator; 2, EVV-34; and 3, cast TNT.

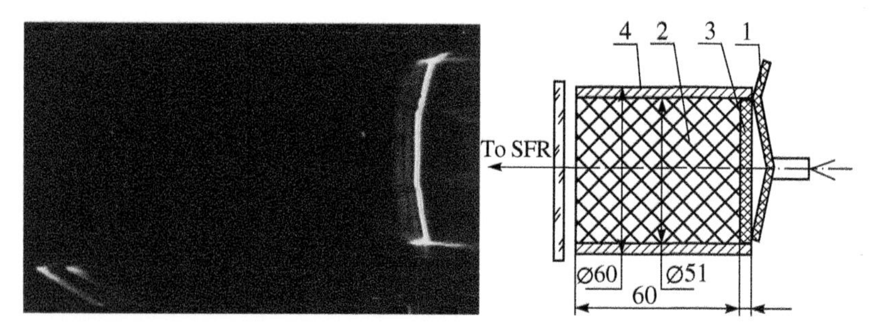

Figure 4.20 Streak camera photograph of the detonation front shape: 1, plane wave generator; 2, cast TNT; 3, EVV-34; and 4, brass shell.

Figure 4.22. One may clearly observe perturbations on the front that can be related to the influence of the outrunning wave in the silicon carbide shell.

The following conclusions can be drawn based on experimental observations. In all experiments without shells, when a plane wave generator was used, classical symmetric flat detonation fronts were observed with a distortion of the front at the periphery of the charge due to the influence of lateral rarefaction waves. In all experiments with charges surrounded by silicon carbide shells, the perturbations of the detonation front shape were observed. This is apparently related to the influence of the outrunning wave propagating in the shell on the process of detonation. Comparison of the results observed in the case of charges with shells of different lengths (see Figures 4.17 and 4.18) shows that the process of detonation propagation is nonstationary. The perturbations of the

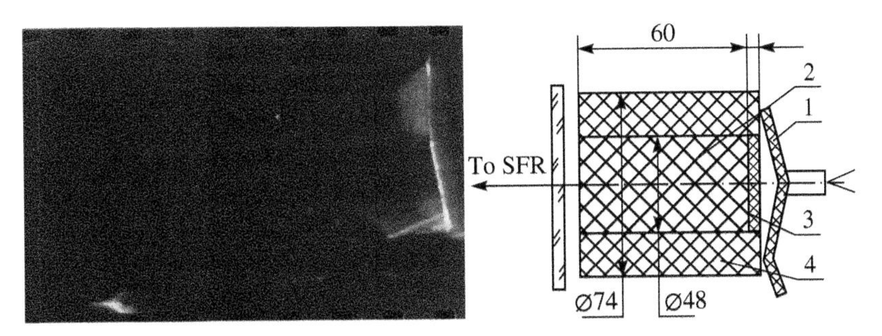

Figure 4.21 Streak camera photograph of detonation the front shape: 1, plane wave generator; 2, cast TNT; 3, EVV-34; and 4, silicon carbide shell.

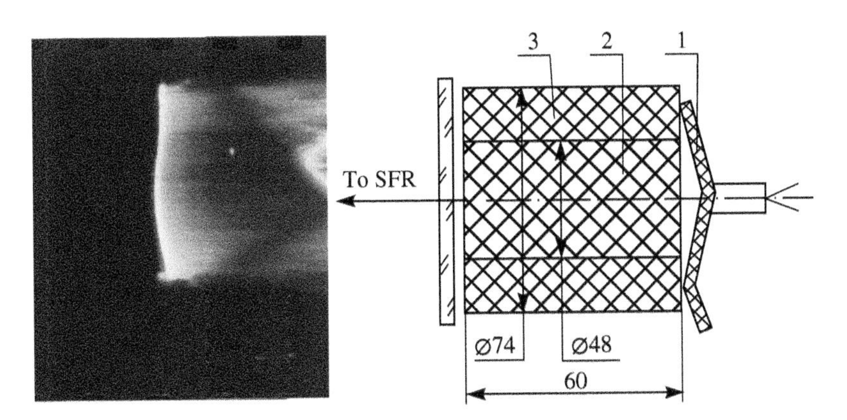

Figure 4.22 Streak camera photograph of detonation front shape: 1, plane wave generator; 2, TGA-40/40/20; and 3, silicon carbide shell.

front in charges embedded in the brass shells were most likely caused by the channel effect due to the presence of gaps (0.1–0.2 mm) between the charge and the shell that appeared due to the shrinkage during casting.

To exclude the influence of the channel effect and possible defects in the structure of the cast HE charges, several additional experiments were performed [9]. The purpose of these experiments was to study the effect of cylindrical shells of brass and silicon carbide on the detonation velocity and the geometry of the detonation wave front in the LVV-11-1 explosive. Elastic HE LVV-11-1 is an explosive based on RDX with the density of 1.42 g/cm^3 and the detonation velocity of 7500 m/s. It is manufactured in the shape of strips with a 4 × 100 mm cross section and a length of 1800 mm. The charges were produced by cutting 48 mm disks out of such a strip. These elements were stacked on top of each other inside a shell of the same diameter. The uniformity of packing in the shell was ensured by using a wooden punch. To exclude the channel effect, the side faces of LVB-11-1 and the inner surface of the shell were smeared with grease. The experimental assembly for measuring the detonation front shape is shown in Figure 4.23.

An ATED-15 electric detonator was used to initiate the detonation. The detonation velocity was measured by electrocontact gauges. To record the process of detonation front exiting at the end of the experimental assembly, a high-speed SFR-2M camera operating in streak mode was used. The following main parameters of registration were used: focal length was 750 mm, rotation speed of the mirror was 60 000 rpm, the recording speed of the film was 3.00 mm/μs, and the distance to the object was 6 m.

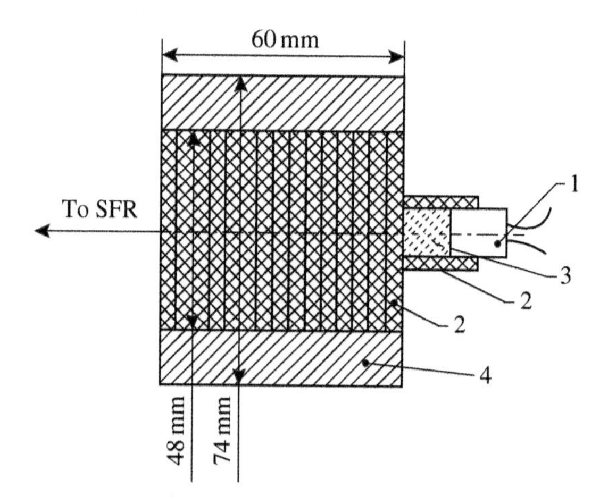

Figure 4.23 Experimental assembly: 1, electric detonator ATED-15; 2, LVV-11-1; 3, RDX; and 4, silicon carbide/brass shell.

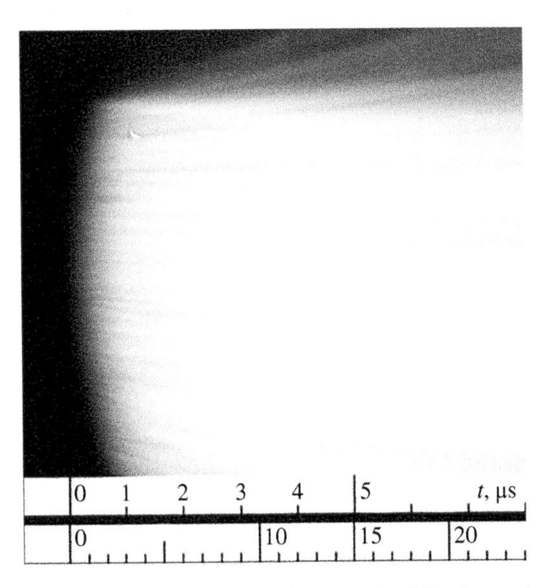

Figure 4.24 Streak camera photograph of the detonation front shape in the experiment with a brass shell.

Figures 4.24 and 4.25 show streak camera photographs of the detonation front output to the rear ends of experimental setups containing brass and ceramic shells, respectively. They demonstrate a significant difference in the wave patterns at the explosive/shell interface. For a high-modulus shell, one may observe a more pronounced glow and the subsequent dynamics of the detonation product glowing.

Additional experiments to measure the average detonation velocity in explosives without shells, as well as in explosives placed in steel, brass, and ceramic shells, have shown that with an accuracy of up to 1.5% the detonation velocity does not depend on the presence of a shell or its material and is equal to 7600 m/s.

4.2 Numerical Analysis of the Influence of Shells on the Detonation Process

To analyze the features of detonation processes in charges placed in ceramic shells, we performed numerical simulations [10–13] closely corresponding to experimental setups. The simulations were carried out using the ANSYS AUTODYN software. The initial configuration of the computational area is shown in Figure 4.26. The simulation was performed in an axisymmetric Lagrangian formulation. The lower

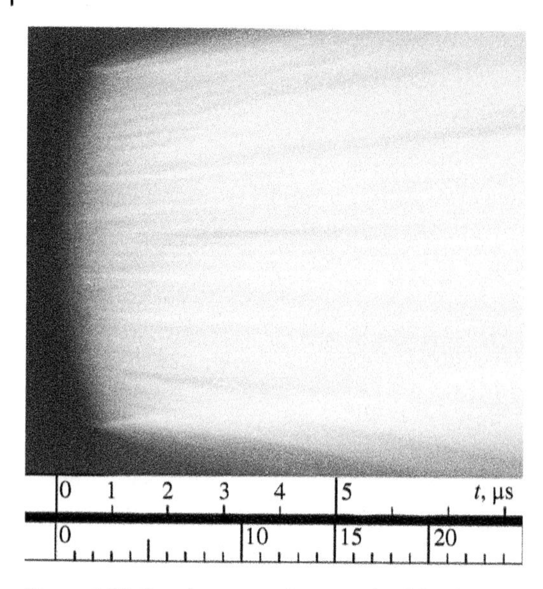

Figure 4.25 Streak camera photograph of the detonation front shape in the experiment with a ceramic shell.

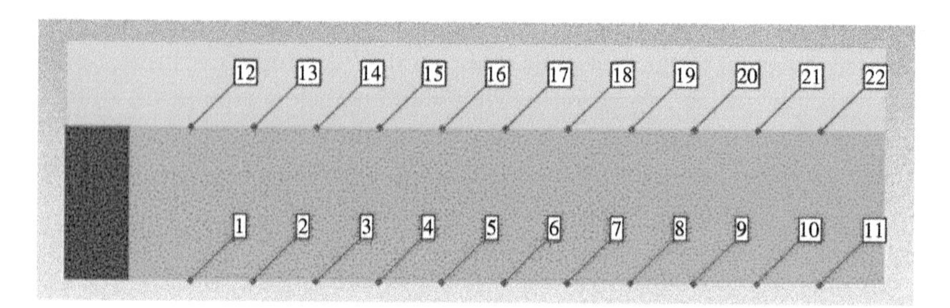

Figure 4.26 Initial configuration of the calculation area: red is the detonation initiation line, blue is the detonator (Composition B $\rho = 1.72$ g/cm^3, JWL), green is the main HE charge (Composition B, $\rho = 1.63$ g/cm^3, Lee–Tarver kinetics), and blue is the shell (copper/silicon carbide); 1–22 are Lagrangian gauges. (*Source:* From Balagansky et al. [13]. Reprinted with permission of Pleiades Publishing, Ltd.)

boundary of the computational area corresponded to the axis of symmetry. The spatial resolution was 10 cells/1 mm. The total physical dimensions of the region are 13 × 3.7 cm, i.e. its length was more than two times the length of previously described experimental setups. The red line in the figure corresponds to the plane of detonation initiation, the blue area is the intermediate charge made of Composition B explosive with the density $\rho = 1.72$ g/cm^3, the green region is the main charge (Composition B with $\rho = 1.63$ g/cm^3), and blue area represents the shell.

The Johns-Wilkins-Lee (JWL) equation of state was used to describe the behavior of Composition B, and the detonation transition was modeled by the Lee–Tarver kinetics. The numbers indicate the location of "Lagrangian gauges" – moving points, in which the change in process parameters with time was recorded. The method of selecting the parameters defining the governing relationships is described earlier. Calculations were performed for shells of silicon carbide and copper that have a similar acoustic impedance.

Figure 4.27 shows the calculated flow fields and pressure contours at the time $t = 15.75$ μs for the copper (a) and the ceramic (b) shell. With the passage of time, quantitative and qualitative differences in wave patterns both in detonation products and in shells increase. The circulation of compression and rarefaction waves is clearly revealed in the expanding copper shell. This behavior also appears in detonation products. In the ceramic shell, the compression wave moves noticeably ahead of the detonation front, and the shell itself is rapidly destroyed.

(a)

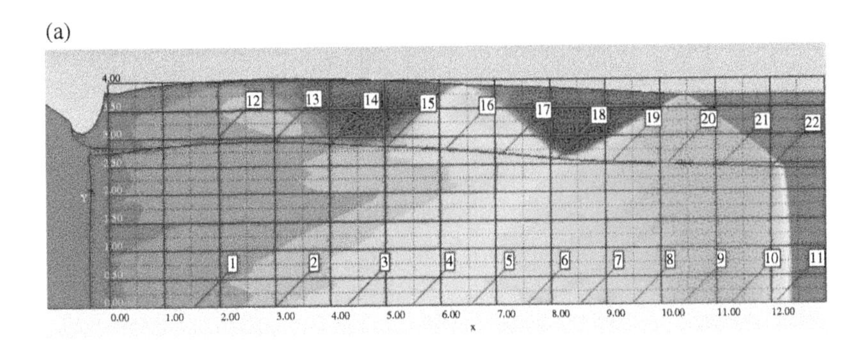

(b)

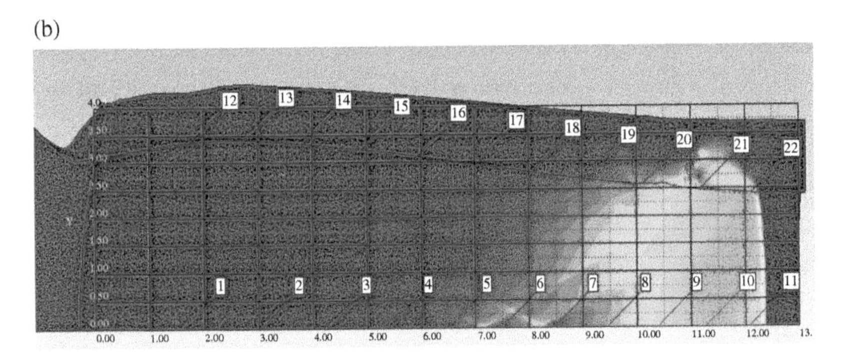

Figure 4.27 Flow fields and pressure contours at time $t = 15.75$ μs for the copper (a) and the ceramic (b) shell. (*Source:* From Balagansky et al. [13]. Reprinted with permission of Pleiades Publishing, Ltd.)

Figure 4.28 shows the spatial distribution of pressure near the detonation front between coordinates $x_1 = 11$ cm and $x_2 = 13$ cm at equivalent times for the copper shell (top graph) and the ceramic shell (bottom graph).

The graphs show a significant decrease in pressure and blurring of the wave front for the silicon carbide shell in comparison with the copper shell. This indicates a significant desensitization of the explosive when it is loaded by a leading wave in the ceramic shell [14–17].

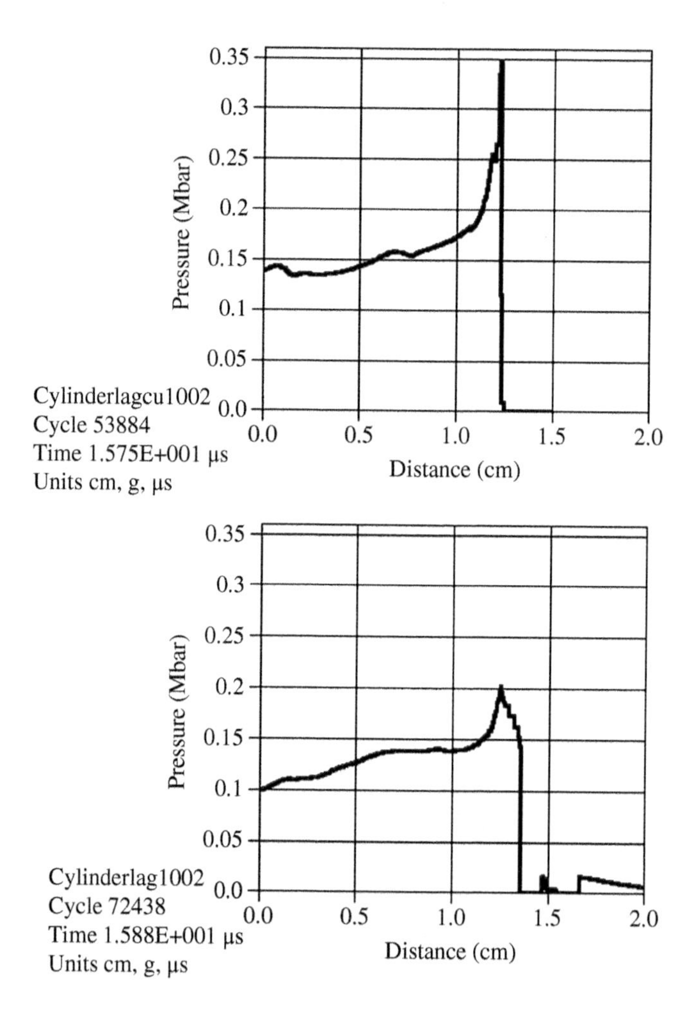

Figure 4.28 Pressure profiles near the front at the interface at equivalent times for the copper (top) and the ceramic (bottom) shell. (*Source:* From Balagansky et al. [13]. Reprinted with permission of Pleiades Publishing, Ltd.)

Figures 4.29 and 4.30 show the dependences of pressure and particle velocity against time in explosive charges at the interface with the shells recorded by "Lagrangian gauges" for simulations with the copper shell (top graph) and the ceramic shell (bottom graph).

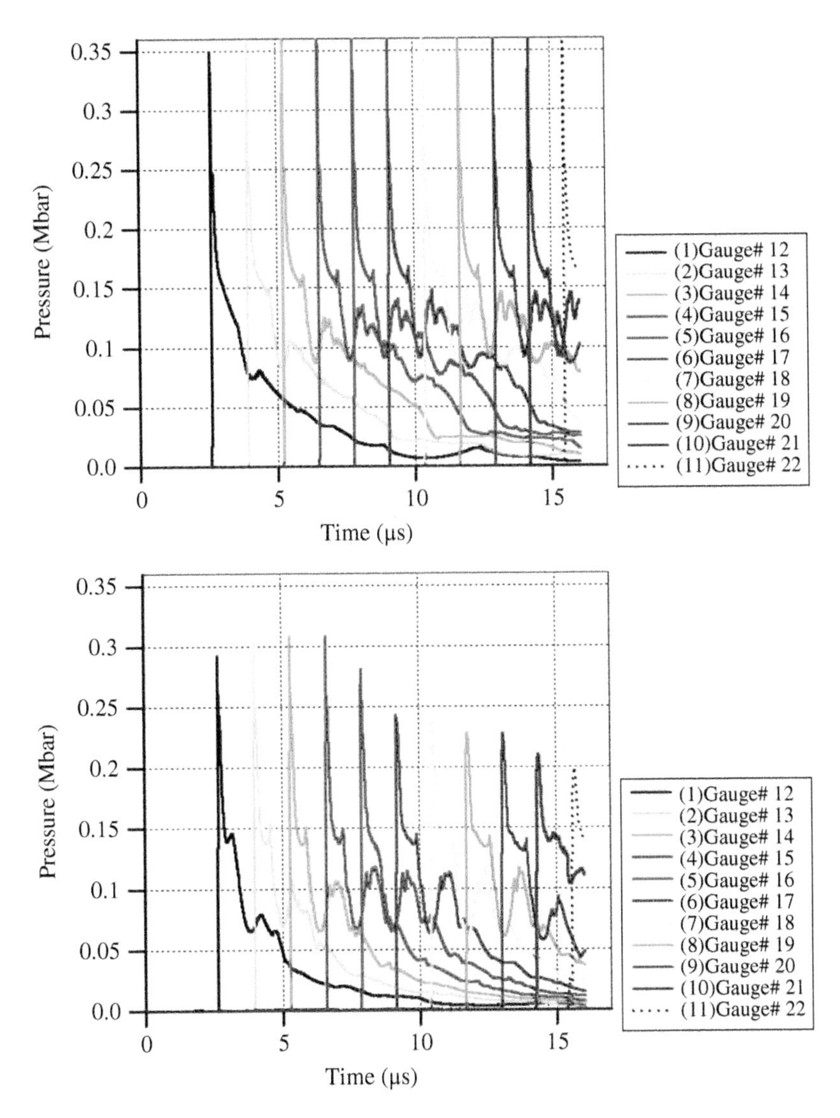

Figure 4.29 Graphs of pressure in the Lagrangian points 12–22 at the interface with the copper (top) and the ceramic (bottom) shell. (*Source:* From Balagansky et al. [13]. Reprinted with permission of Pleiades Publishing, Ltd.)

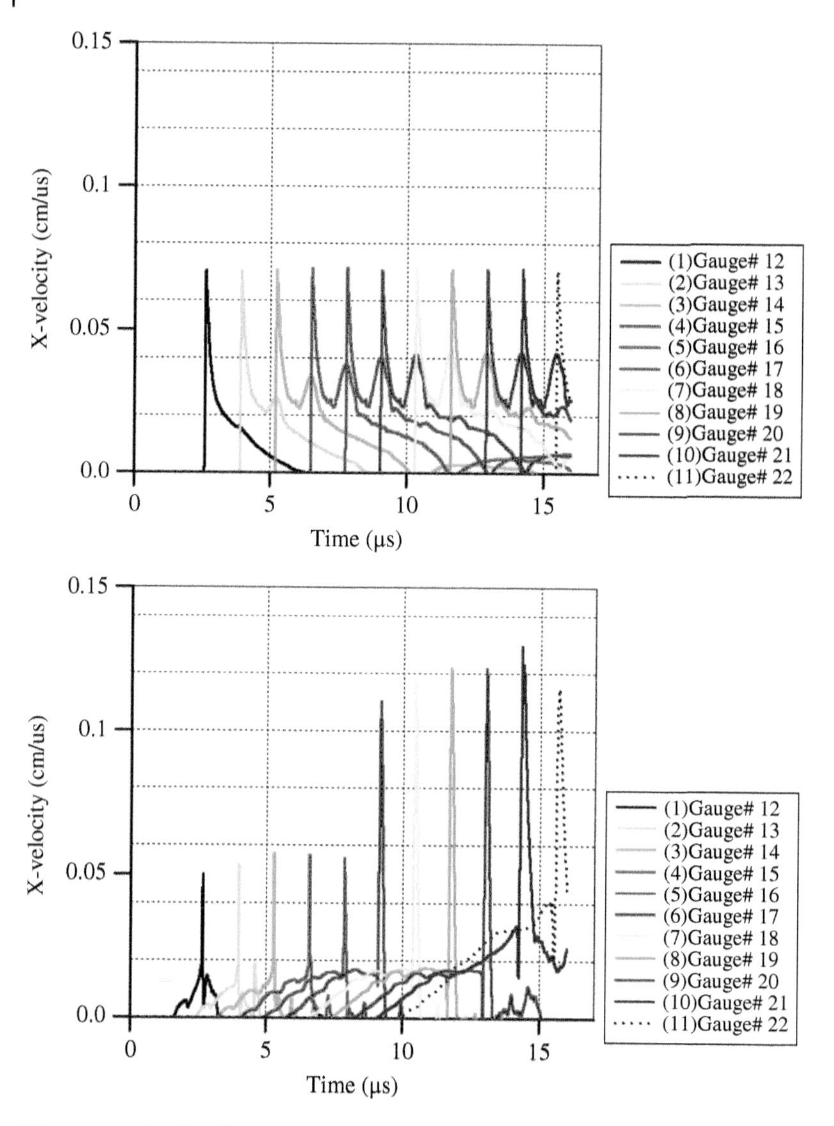

Figure 4.30 Graphs of particle velocity in the Lagrangian points 12–22 at the interface with the copper (top) and the ceramic (bottom) shell. (*Source:* From Balagansky et al. [13]. Reprinted with permission of Pleiades Publishing, Ltd.)

For the charge in a copper shell, the pressure and particle velocity at the detonation front remain constant as it propagates, which means that the detonation process is stationary. For the charge in a silicon carbide shell, a continuous decrease in pressure and an increase in particle velocity are observed, which is a sign of a not fully compressed detonation.

Figure 4.31 Flow fields and pressure contours in charges with copper (left graph) and ceramic (right graph) shells at $t = 8$ μs (zoom).

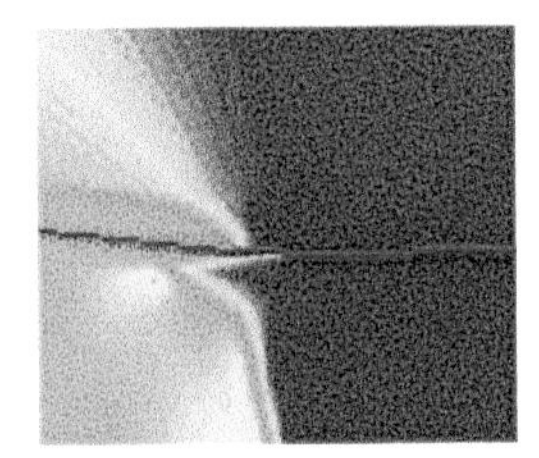

Figure 4.32 Flow fields and pressure contours in charges with a copper (left) and ceramic (right) shell, $t = 15.75$ μs (zoom). (*Source:* From Balagansky et al. [13]. Reprinted with permission of Pleiades Publishing, Ltd.)

Figure 4.31 shows the calculated flow fields and pressure contours in charges with copper (left graph) and ceramic (right graph) shells at the time $t = 8$ μs. The graphs show the cropped regions of the interaction of the detonation front with the shells. While for the copper shell, a classical break of the "detonation front – shock wave" line is observed, the shock wave in the ceramic shell outruns the detonation front.

At time $t = 15.75$ μs, the flow picture in a charge with a ceramic shell radically changes (Figure 4.32, right graph). Directly on the boundary, there is a local advance of detonation front, while at some distance from the boundary there is a local delay of the detonation front. In such case, a perturbation nucleates and its development distorts the shape of the front. The flow picture for the charge in the copper shell remained practically unchanged.

Figures 4.33 and 4.34 show flow fields and contours of the decomposition degree of explosive at the same moments of time (8 and 15.75 μs) as in Figures 4.31 and 4.32 near the interaction zone of the detonation front with the copper shell (left graphs) and the ceramic shell (right graphs). The data presented here confirm the previously drawn conclusions about the peculiarities of the processes in charges with copper and ceramic shells.

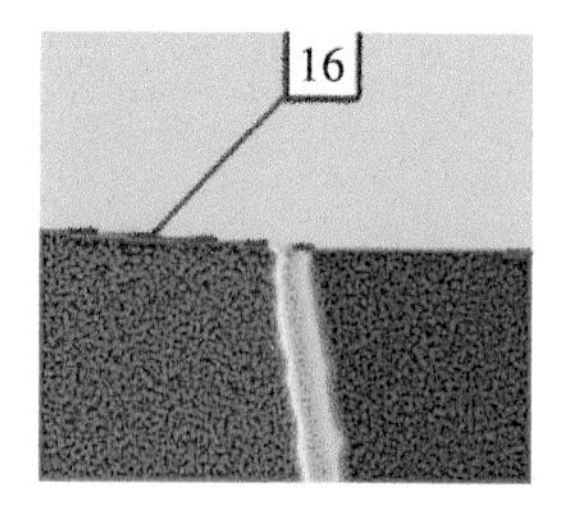

Figure 4.33 Flow fields and decomposition degree contours in the charges with copper (left graph) and ceramic (right graph) shells for the time $t = 8\ \mu s$ (zoom).

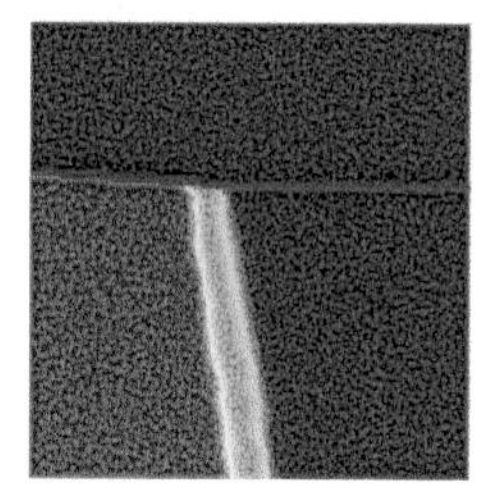

Figure 4.34 Flow fields and decomposition degree contours in the charges with copper (left graph) and ceramic (right graph) shells for the time $t = 15.75\ \mu s$ (zoom).

The results show that while in a charge with a copper shell, the detonation process remains stationary, in a charge with a ceramic shell there is a substantially nonstationary process at the interface, in which the mutual location of the shock wave (compression wave) in the shell wall and detonation front in explosive periodically changes, which causes a subsequent distortion of the detonation front shape. This is accompanied by a significant slowdown in the rate of explosive decomposition directly on the boundary with the ceramic shell.

Similar results were obtained, for example, in the study of the detonation wave propagation in charges of emulsion explosives encased in shells from materials with a high sound velocity, as shown in Figure 4.35 (isolines of pressure for the charge with a steel shell are shown on the left and for one with an aluminum shell on the right) [18].

The duration of the detonation transformation at the locations of the gauges is shown in Figure 4.36. The top graph corresponds to the charge with a copper shell, while the bottom one is for the charge with a ceramic shell.

On the axis of symmetry, the duration of the detonation transformation remains practically constant throughout the process for both copper and ceramic shells. At the same time, at the interface, the duration of the detonation transformation in a charge with a ceramic shell increases with

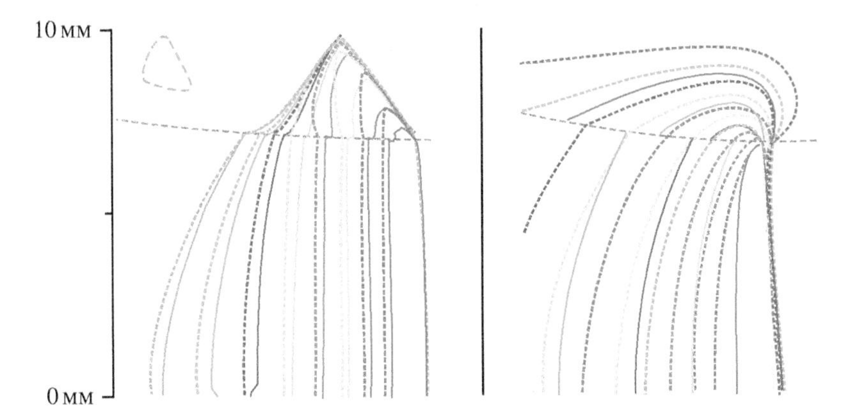

Figure 4.35 Isolines of pressures in the charge of emulsion explosives with steel (left) and aluminum (right) shells. (*Source:* From Schoch et al. [18]. Reprinted with permission of AIP Publishing.)

time and reaches 0.44 μs at the time $t = 15.75$ μs. This exceeds the maximum time of explosive transformation at the interface of the charge with the copper shell by more than four times.

This happens due to the desensitization of explosives in the area adjacent to the shell. At the same time, the reaction at the boundary begins earlier than in regions that are at some distance from the boundary.

Figure 4.37 shows the detonation front profiles on the axis of symmetry of explosive charge at the time $t = 15.75$ μs.

Behind the detonation front in the explosive charge with a ceramic shell, one may observe a prolonged region with an almost constant pressure, which is close to the pressure at the Chapman–Jouguet point. The decomposition time of the explosive is practically the same as for the charge in a copper shell. This means that the formation of this region is caused by the compression wave coming from the periphery of the charge (that is, from the shell side).

Figures 4.38–4.43 show the successive stages of the formation and development of perturbations of the detonation front for a charge in a ceramic shell. For clarity, only four isolines of pressures are shown having the lower boundary of 0.215 Mbar and the upper boundary of 0.230 Mbar.

4.3 Summary

The influence of the shell material (copper and silicon carbide) on the detonation process of a cylindrical explosive charge was experimentally and numerically investigated. Despite practically the same dynamic

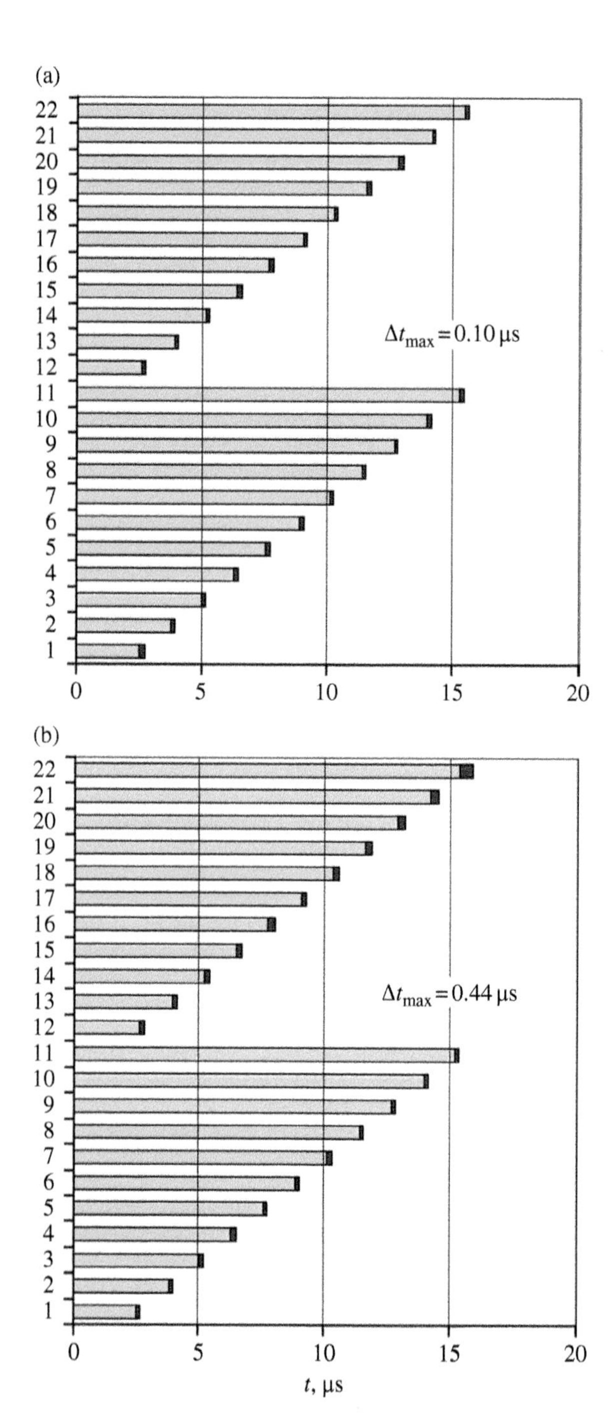

Figure 4.36 Duration of detonation transformation in places of gauges locations in microseconds: (a), copper; (b), ceramics. The vertical axis shows the numbers of gauges, and dark rectangles indicate the time of complete conversion of explosives. (*Source:* From Balagansky et al. [13]. Reprinted with permission of Pleiades Publishing, Ltd.)

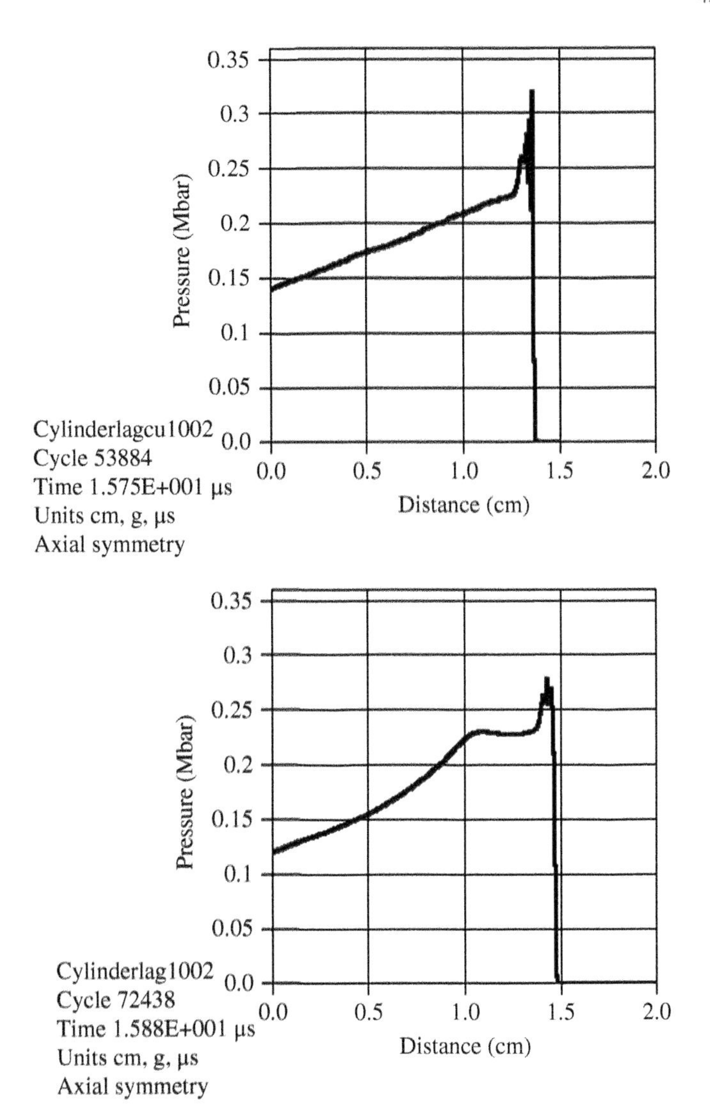

Figure 4.37 Pressure profiles near the front on the symmetry axis of the HE charge: copper (top) and ceramics (bottom). (*Source:* From Balagansky et al. [13]. Reprinted with permission of Pleiades Publishing, Ltd.)

rigidity of shell materials, there are significant differences in wave processes both in detonation products and in shells, which is caused by the differences in the sound velocities of copper and silicon carbide and by the rapid destruction of the ceramic under explosive loading.

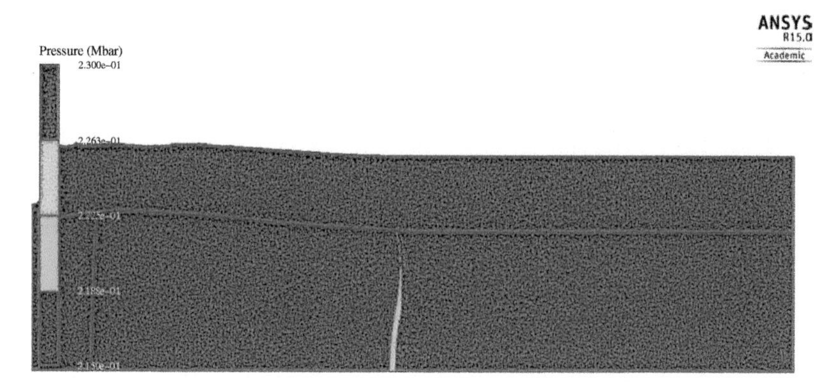

Figure 4.38 Transverse waves in the explosive charge near the detonation front at time $t = 8$ μs.

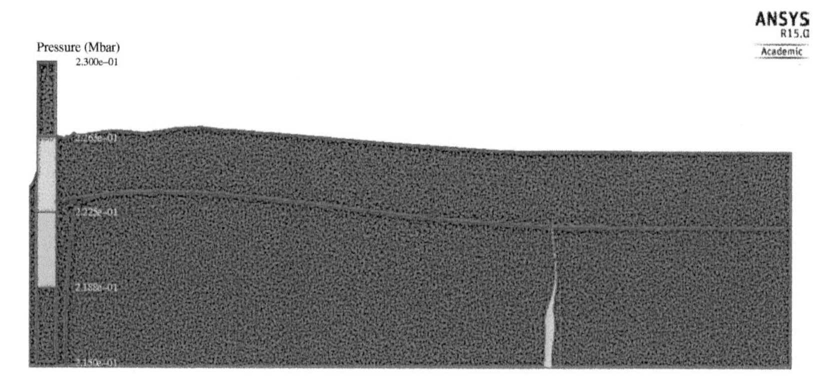

Figure 4.39 Transverse waves in the explosive charge near the detonation front at time $t = 11.15$ μs.

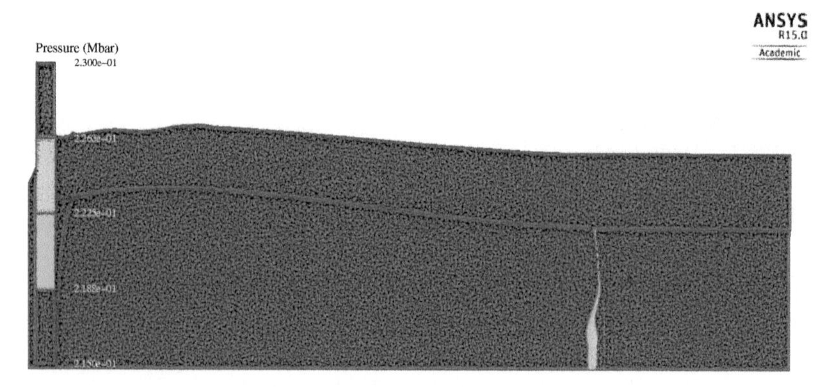

Figure 4.40 Transverse waves in the explosive charge near the detonation front at time $t = 12.5$ μs.

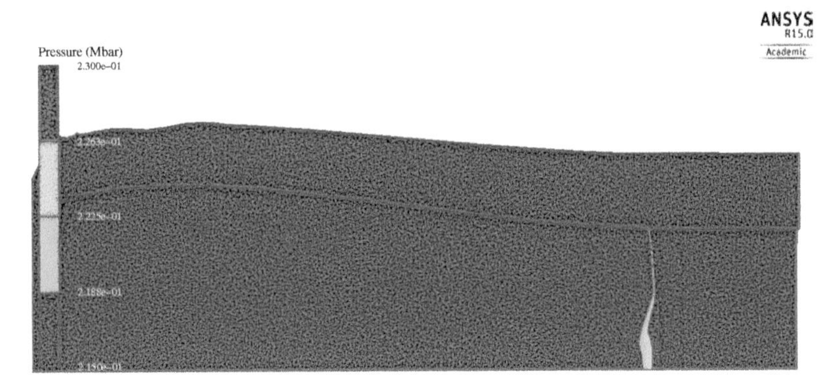

Figure 4.41 Transverse waves in the explosive charge near the detonation front at time $t = 13.5$ μs.

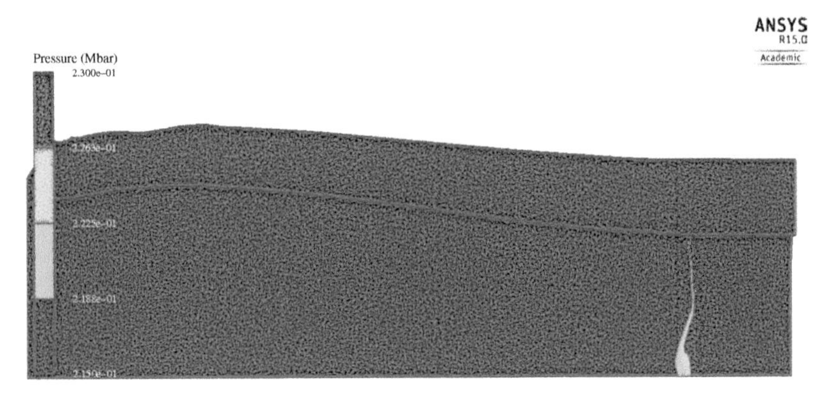

Figure 4.42 Transverse waves in the explosive charge near the detonation front at time $t = 14.5$ μs.

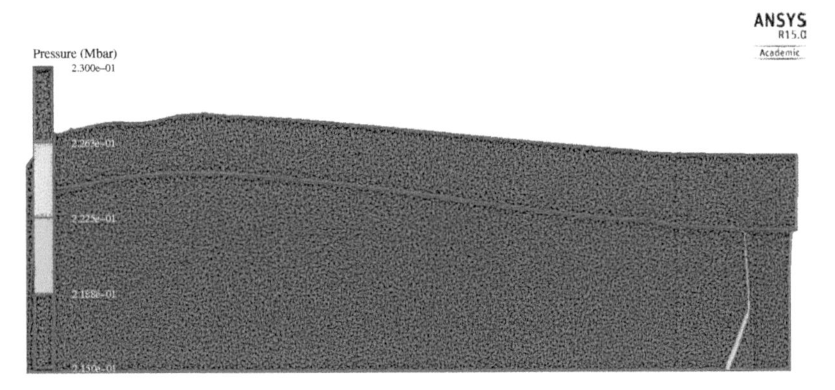

Figure 4.43 Transverse waves in the explosive charge near the detonation front at time $t = 15.75$ μs.

The peculiarities of the wave propagation at the explosive/ceramic interface associated with the desensitization of the explosive due to its loading by the leading wave at the shell side were revealed. In this case, this leads to a pressure decrease, smearing of the detonation front, and an increase in the particle velocity, which is typical for an incompletely compressed detonation. Throughout the whole process, the explosive decomposition time at the interface between the explosive and the shell continuously increases up to four times. Thus, near the boundary, a nonstationary detonation regime is realized.

On the symmetry axis, behind the detonation front of the explosive charge in the ceramic shell, there is a prolonged region with an almost constant pressure close to the pressure value at the Chapman–Jouguet point. The calculation results indicate that the time of the explosive decomposition on the symmetry axis is practically the same as for the charge in a copper shell.

A mechanism for transferring perturbations from the periphery to the symmetry axis of the explosive charge was revealed. Formation of this region is caused by the transverse compression waves propagating from the charge periphery to the symmetry axis immediately behind the detonation front. When converging to the axis of symmetry, they lead to an increase in the duration of the pressure action behind the detonation front.

In general, the main features of the process are well reproduced in numerical modeling. Possible differences between the experimental and calculated data can be attributed to the effect of artificial viscosity used in the simulation on the processes associated with the kinetic instability of the detonation front.

References

1 Hornberg, H. (1986). Determination of fume state parameters from expansion measurements of metal tubes. *Propellants, Explosives, Pyrotechnics* 11: 23–31.

2 Hornberg, H. and Volk, E. (1989). The cylinder test in the context of physical detonation measurement methods. *Propellants, Explosives. Pyrotechnics* 14: 199–211.

3 Sanchidrián, J.A. and López, L.M. (2006). Calculation of the energy of explosives with a partial reaction model. Comparison with cylinder test data. *Propellants, Explosives, Pyrotechnics* 31 (1): 25–32.

4 Odintsov, V.A. and Chudov, L.A. (1975). Expansion and destruction of shells under the action of detonation products. In: *Problems of Dynamics of Elastoplastic Media* (ed. G.S. Shapiro). Moscow: Mir [in Russian].

5 Balagansky, I.A., Berdnik, V.P., Kulikova, I.V. et al. (1991). Features of detonation processes in HE charges that contact with high modulus ceramics. *Annotations of the reports of the 7th All-Union Congress on Theoretical and Applied Mechanics*. Moscow, USSR (15–21 August 1991). Moscow: Moscow State University. [in Russian].
6 Balagansky, I.A., Kobylkin, I.F., Razorenov, S.V. et al. (1991). Effect of a silicon carbide shell on detonation parameters in high explosives. *Proceedings of 5 All-Union Conference on Detonation*. Krasnoyarsk, USSR (5–12 August 1991). Krasnoyarsk: Russian Academy of Science. [in Russian].
7 Balagansky, I.A., Razorenov, S.V., Utkin, A.V. (1993). Detonation parameters of condensed high explosive charges with long ceramic elements. *Proceedings of the 10th International Detonation Symposium*, Boston, USA (12–16 July 1993). Arlington: Office of Naval Research.
8 Balagansky, I.A., Balagansky, A.I., Kobilkin, I.F. et al. (2005). Influence of high explosive charge shell on detonation front shape. *Proceedings of International conference 'VIII Zababakhin Scientific readings'*, Snezhinsk, Russia (5–9 September 2005). Snezhinsk: RFNC. [in Russian].
9 Balagansky, I.A., Vinogradov, A.V., Merzhievsky, L.A. et al. (2016). Analysis of shell material influence on detonation process in high explosive charge. *Key Engineering Materials* 715: 27–32.
10 Balagansky, I.A., Vinogradov, A.V., and Merzhievsky, L.A. (2015). Influence of ceramic shell on detonation process. *Proceedings of 42 scientific and technical conference 'Designing of systems'*, Moscow, Russia (4–6 February 2015). Moscow: Bauman Moscow State Technical University. [in Russian].
11 Balagansky, I.A., Vinogradov, A.V., and Merzhievsky, L.A. (2015). Numerical analysis of the influence of shell material on the detonation process. *Abstracts of Lavrentyev's Readings on Mathematics, Mechanics and Physics*, Novosibirsk, Russia) (7–11 September 2015). Novosibirsk: Institute of Hydrodynamics, SB RAS. [in Russian].
12 Balagansky, I.A., Vinogradov, A.V., Merzhievsky, L.A. et al. (2016). Analysis of shell material influence on detonation process in high explosive charge. *Proceedings of Fifth International Symposium on Explosion, Shock Wave and High-Strain-Rate Phenomena*, Beijing, China (25–28 September 2016). Beijing: Beijing Institute of Technology.
13 Balagansky, I.A., Vinogradov, A.V., Merzhievsky, L.A. et al. (2018). Analysis of the influence of the shell material on the detonation process in high explosive charge. *Combustion, Explosion and Shock Waves* 54 (4): 502–510.
14 Jacobs, S.J. (1960). Non-steady detonation. *Proceedings of 3rd Symposium on Detonation*, Princeton, USA (26–28 September 1960). Arlington: Office of Naval Research.

15 Campbell, A.W., Davis, W.C., Ramsay, J.B. et al. (1961). Shock initiation of solid explosives. *Physics of Fluids* 4: 511–521.
16 Campbell, A.W. and Travis, J.R. (1985). The shock desensitization of PBX-9404 and Composition B-3. *Proceedings of 8th Symposium on Detonation*, Albuquerque, USA (15–19 July 1985). Silver Spring: Naval Surface Weapons Center.
17 Davis, W.C. (2010). Shock desensitizing of solid explosive. *Proceedings of 14th International Detonation Symposium*, Coeur d'Alene, USA (11–16 April 2010). Arlington: Office of Naval Research.
18 Schoch, S., Nikolaos, N., and Lee, B.J. (2013). The propagation of detonation waves in non-ideal condensed-phase explosives confined be high sound-speed materials. *Physics of Fluids* 25: 086102.

5

Hypervelocity of Shaped Charge Jets

The devices for hypervelocity acceleration of impactors are widely used in the research and development of protection of spacecraft against meteors and orbital debris. As a rule, these devices are based on the phenomenon of cumulation. Cumulation may be described as the phenomenon of explosion energy concentration in a given direction, which leads to a significant local increase in its destructive effect, which is usually achieved by the cumulative jet formed during the collapse of a specially shaped liner by an explosive charge. As a rule, ductile materials such as copper are used for these liners, which form continuous jets providing a high penetrating effect.

Apparently, the first publication devoted to the study of hypersonic cumulative jets was [1], in which beryllium liners were used. Beryllium has a high sound velocity, and it is known that the maximum speed of a cumulative jet V_j capable of penetrating barriers of considerable thickness cannot exceed the value [2, 3]

$$V_j \leq 2.41 C_b,$$

where C_b is the volume velocity of sound in the liner material.

This limit is valid for jets capable of penetrating a barrier. It should be noted that this limit will not necessarily be reached in every shaped charge. In HE charges with beryllium liners, gaseous jets were formed, having a very high velocity. However, their density was so low that they could only be observed in a vacuum. The highest velocities of such jets for beryllium liners were approximately 90 km/s. But such a gaseous jet cannot produce any noticeable damaging effect.

One of the types of shaped charge accelerators is the gas-cumulative accelerator described in [4]. In this case, an explosive charge is shaped as a tube inside which a cumulative jet of detonation products is formed, which propagates inside the tube at a velocity much higher than the

Explosion Systems with Inert High-Modulus Components: Increasing the Efficiency of Blast Technologies and Their Applications, First Edition. Igor A. Balagansky, Anatoliy A. Bataev, and Ivan A. Bataev.

detonation velocity. This jet can accelerate a body (usually a sphere) to velocities of the orbital range (8–14 km/s).

It is known [5–7] that the formation of cumulative jets occurs if the point of collision during the collapse of the cumulative liner moves with a velocity V_k not exceeding the volume velocity of sound in the material of the liner C_b:

$$V_k \le C_b.$$

To increase the cumulative jet velocity, it is possible to form cumulative jets from cylindrical liners (tubes) [6, 7]. Since for a cylindrical liner, the collision point velocity is equal to the detonation velocity D in the explosive charge,

$$V_k = D,$$

and a necessary condition for the formation of jets for cylindrical liners is

$$C_b \ge D.$$

Thus, to design cylindrical liners for high-velocity jets, one may use high-modulus ceramics having a sound velocity exceeding the detonation velocity of explosives. However, such materials do not form continuous jets.

5.1 Experimental Investigation of Ceramic Tube Collapse by Detonation Products

In the first series of the experiments [8–10], we used the tubes of pure alumina (commercially available as the "Policor" material), manufactured by extrusion from a melt. The density of the material was 3.96 g/cm^3, which makes it possible to conclude that it was the α-modification of aluminum oxide (theoretical density of which is 3.99 g/cm^3), the velocity of sound, in this case is approximately 10 km/s. The length of the tubes was 103 mm, the outer diameter was 9 mm, and the inner diameter was 7 mm. The tubes were positioned along the axis of symmetry of a cast TG-40 charge (ρ_0 = 1.69 g/cm^3). In all experiments, the detonation was initiated by plane wave generators. The results were recorded using the SFR-2 M and ZHLV-2 cameras in streak and frame modes, respectively.

The scheme of the shaped charge used to study ceramic cumulative jets is shown in Figure 5.1. The successive stages of the formation and development of a high-speed cumulative jet recorded by the ZHLV-2 camera in the frame mode with a time step of 2.66 μs are shown in Figure 5.2 (a reference grid with 25 × 25 mm cells was used to simplify measurements). The photographs show that the ceramic jet manifests itself as a brightly

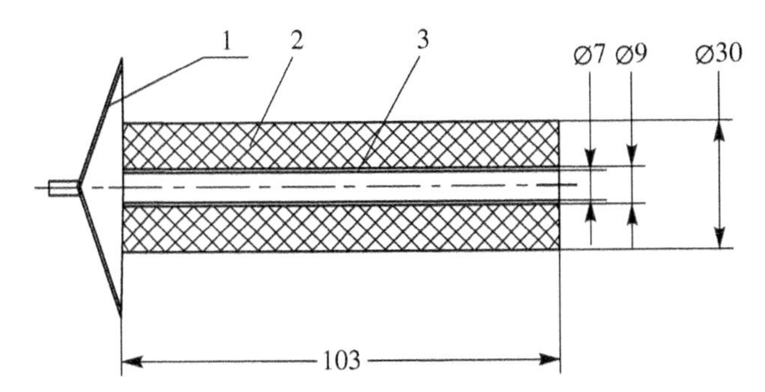

Figure 5.1 Experimental assembly scheme: 1, plane wave generator; 2, HE charge; and 3, ceramic tube. (*Source:* From Balagansky et al. [9]. Reprinted with permission of Elsevier.)

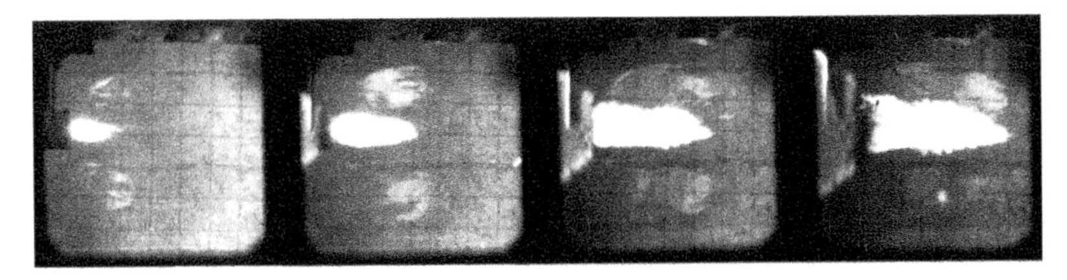

Figure 5.2 Cumulative jet movement in the air. Grid cell size is 25 × 25 mm. (*Source:* From Balagansky et al. [9]. Reprinted with permission of Elsevier.)

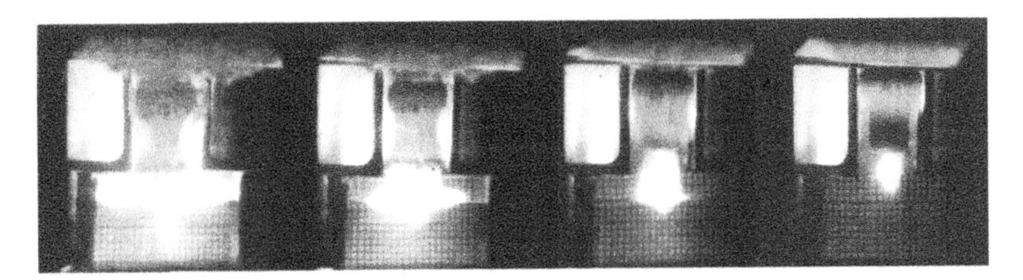

Figure 5.3 The process of penetration of the formed cumulative jet (particle flow) into the water. (*Source:* From Balagansky et al. [9]. Reprinted with permission of Elsevier.)

glowing stream of particles, which move much faster than the propagation of detonation in the explosive charge.

Figure 5.3 shows the process of penetration of the cumulative jet (particle flow) into water (the sequential time order of frames is from right to left, grid cell size is 5 × 5 mm).

Registering the movement of a jet in the air (Figure 5.4) was performed using the SFR-2 M camera in the streak mode. The writing speed on the

Figure 5.4 Streak camera photograph of cumulative jet movement in the air. (*Source:* From Balagansky et al. [9]. Reprinted with permission of Elsevier.)

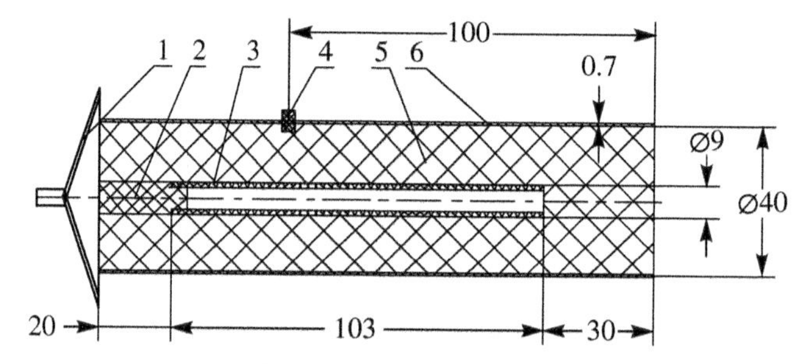

Figure 5.5 The scheme of the experiment: 1, plane wave generator; 2, plug; 3, ceramic tube; 4, time marker (EVV-34); 5, HE; and 6, shell (duralumin).

film was 3.75 mm/μs, the optical reduction ratio was $n = 3.48$. The photograph shows two glowing fronts. The first one moves at a speed of 11 km/s, while the second one at a speed of 12.4 km/s. It can be assumed that this feature of the flow is caused by the influence of the leading wave in the ceramic tube on the adjacent layers of explosives or is associated with edge effects.

The next series of experiments were conducted using the setup shown in Figure 5.5. In these experiments, a foam plug (2) was placed in one end of the ceramic tube (3). The cavity of a hollow charge placed into a thin duralumin shell (6) was not through so that upon exiting the tube the jet would immediately hit the explosive.

The streak camera photograph showing the detonation wave exiting to the rear end of the charge is presented in Figure 5.6. The arrow indicates the beginning of the registration – the glow from the time marker (4), the

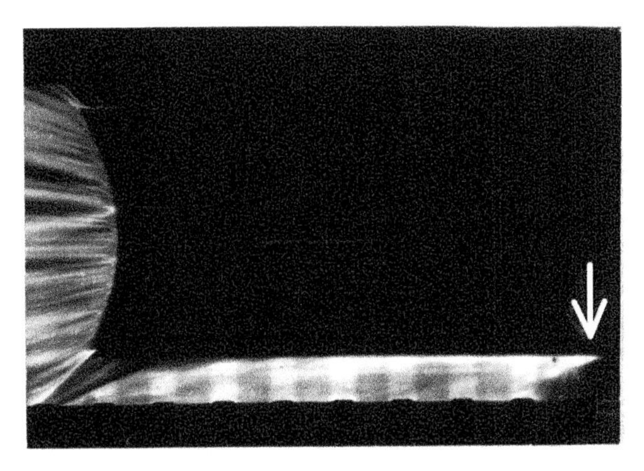

Figure 5.6 Streak camera photograph of the output of the detonation wave to the end of the charge.

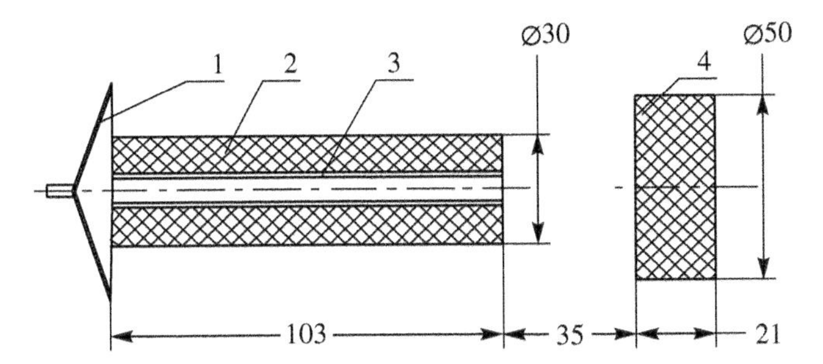

Figure 5.7 Scheme of the experiment on HMX charge initiation: 1, plane wave generator; 2, HE; 3, ceramic tube; and 4, HMX charge.

writing speed was 3.75 mm/μs, the direction of time is from right to left. The detonation on the axis is ahead of the detonation at the edges, which indicates the initiation of the central region of the final part of the charge by a cumulative jet. The average detonation velocity on the path between the flash of the marker and the free surface (100 mm) was about 10 000 m/s. Detonation along the axis outruns the detonation at the periphery by 0.72 μs. This fact indicates the presence of a central initiation of the final portion of the charge and confirms the assumption that the cumulative jet formed because of the collapse of the ceramic tube is capable of initiating detonation in explosive charges. Its initiating ability is confirmed by an experiment conducted according to the setup shown in Figure 5.7. In this case, a charge of phlegmatized HMX with a density of 1.74 g/cm^3 was placed coaxially in front of an explosive charge with a tube (at a distance of 35 mm from its free end). The photograph in

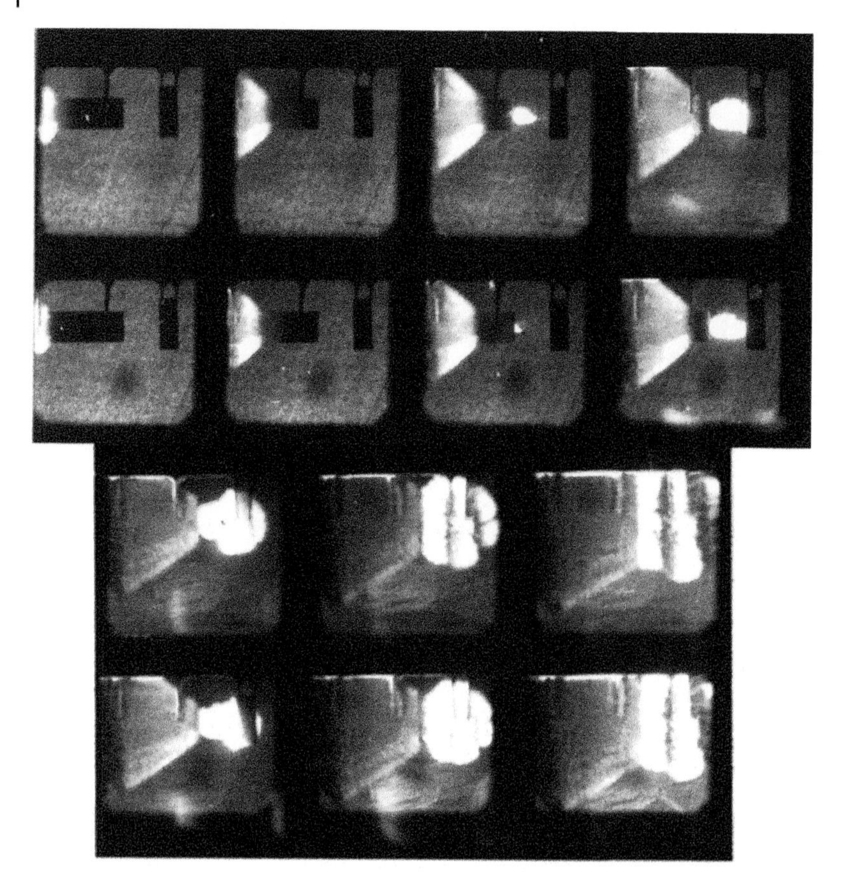

Figure 5.8 Initiation of an HMX charge by the flow of ceramic particles.

Figure 5.8 (time step between frames is 1.33 μs) confirms that, in this case, the jet of the ceramic particles initiates the detonation in the HMX charge.

An interesting effect of the ceramic cumulative jet formation was observed in experiments on shock loading of a silicon carbide ring [11] (Figure 5.9). As follows from the scheme of the experimental setup, the ring was loaded by a shock wave formed at the exit of the detonation wave on the brass plate separating the explosive and the ring.

The process was recorded by a ZHLV-2 camera in the frame mode (Figure 5.10). The time step between the frames was 2.66 μs. In the photographs, one can see a deformation wave that propagates along the ring at a speed of 11–12 km/s, which corresponds to the elastic wave velocity. Behind the deformation wave, a fracture wave is observed. There is no sharp boundary between the elastic wave and the fracture wave.

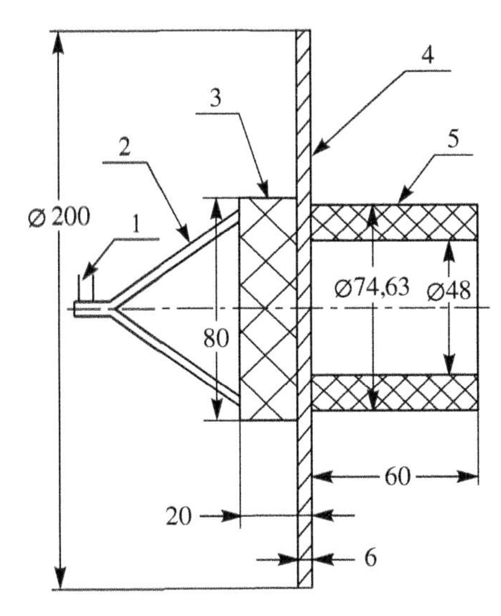

Figure 5.9 The scheme of the experiment: 1, detonator; 2, the cone of EVV-34; 3, pressed RDX; 4, brass disk; and 5, silicon carbide ring.

Figure 5.10 The results of the experiment.

Subsequently, the radial expansion of silicon carbide particles is observed. One may notice a specific feature, which is manifested as a flow of silicon carbide particles moving along the symmetry axis of the ring at a velocity of about 2 km/s. Thus, in the case of shock wave loading of a high-modulus ceramic ring, a cumulative effect is also realized.

5.2 Modeling of Jet Formation Process

Numerical simulation of the jet formation process was performed using several software packages. The ODVAX software for the VAX/VMS system [9, 10], the AUTODYN software in the two-dimensional (2D) formulation [12–14], and the LEGAK method [15, 16] were used.

When the ODVAX software was used, the simulation of explosive collapse process of an alumina tube was conducted in a 2D axisymmetric formulation in the hydrodynamic approximation, taking possible fracture due to the rarefaction waves into account. The simulation setup closely corresponded to the experimental setup shown in Figure 5.1. The behavior of the ceramics was described by the model of an ideal compressible fluid in the following form:

$$p = a^2 \rho \left(\frac{\rho}{\rho_0} - 1 \right) \left[\lambda - (\lambda - 1) \left(\frac{\rho}{\rho_0} \right) \right]^{-2},$$

where ρ_0 is the initial density, ρ is the actual density, and a and λ are the constants in the shock adiabat equation:

$$U_s = a + \lambda U_p.$$

The shock adiabat for alumina has the following form:

$$U_s = 8.00 + 1.5 U_p.$$

The critical stresses of failure were assumed to be −40 GPa. Integration of the continuum mechanics equations was performed using the method of individual particles [17] on a regular rectangular Eulerian mesh with a spatial resolution of 1 cell/mm. On the symmetry axis, a boundary condition for rebound and slip was set. The behavior of explosives was described by a simplified model with constant Chapman–Jouguet parameters at the detonation front. The detonation products were considered to be a polytropic gas. If the pressure in the particle of the detonation products fell below 0.5 GPa or the mass velocity exceeded twice the local sound velocity, then such particles were thrown out from the computation.

The results of the calculations are presented in Figure 5.11. The time points are indicated under each of the figures in microseconds. Figure 5.12 shows the distribution of velocities along the length of the jet and the slug at the time of 15.75 μs. At the initial stage, when the tube collapses, the pressure on the symmetry axis reaches a value of about 900 GPa, and the collapse velocity is about 4 km/s. Subsequently, the pressure at the point of contact oscillates in the range between 80 and 110 GPa. The contact point velocity is equal to the detonation velocity of the explosive charge and is about 8 km/s. The leading part of the stream of ceramic particles moves at a speed of 11 km/s, which is in a very good agreement with experiments. The jet has practically no velocity gradient along its length. According to the kinematic characteristics measured in the experiments, there is a good agreement between the experimental and calculated data.

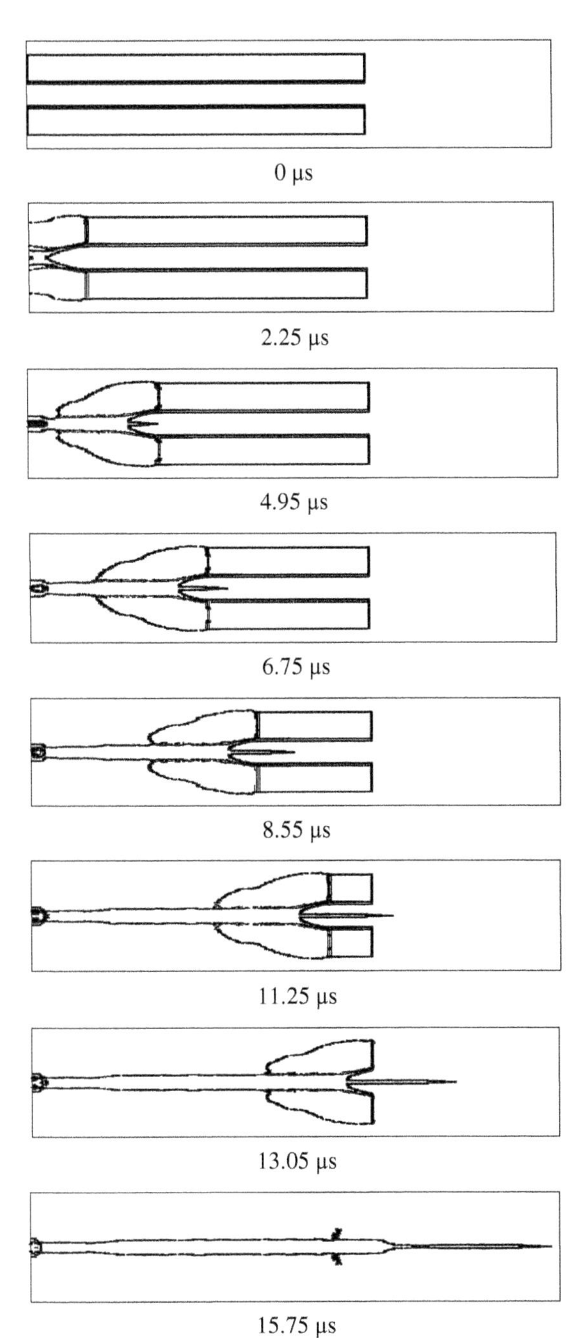

Figure 5.11 Configurations of calculation area at different times. (*Source:* From Balagansky et al. [9]. Reprinted with permission of Elsevier.)

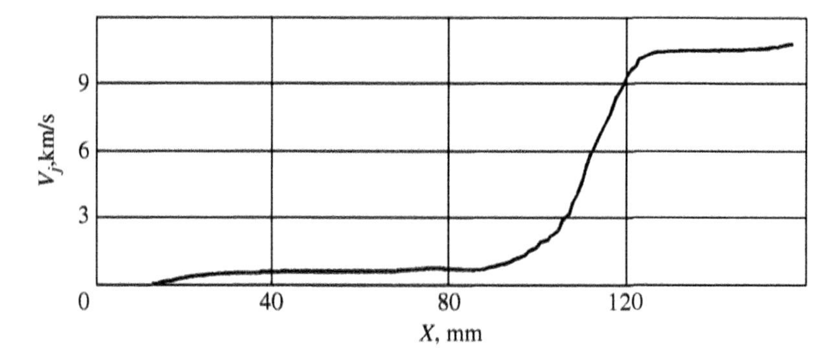

Figure 5.12 The plot of the final jet and slug velocity versus axial distance at $t = 15.75\,\mu s$. (*Source:* From Balagansky et al. [9]. Reprinted with permission of Elsevier.)

More informative was the simulation of the tube collapse process the using the AUTODYN software in the 2D formulation. The choice of models and parameters used in the computation was described earlier. The process of an alumina ceramic tube collapse was simulated in a setup that corresponded to the experiments shown in Figure 5.1. It was assumed that stationary detonation was initiated instantaneously over the entire left end of the charge and Chapman–Jouguet parameters were achieved on its front. The detonation products were described by the JWL equation of state (EOS). For the ceramic tube, two types of the equations of state from the standard AUTODYN material library were used: polynomial approximation and shock adiabat.

Calculations shown in Figure 5.13 indicated that when a ceramic tube collapses, a cumulative jet forms and almost immediately breaks down into separate fragments, which corresponds to experimental observations. The head particles have a velocity that is much higher than the velocity of the bulk of the particles, which also corresponds to the experimental data on the separation of the jet into the high-speed and the main fragments.

An important characteristic that influences the formation of a cumulative jet is the collision angle. In the axisymmetric case, it is the angle between the generatrix of the collapsing shell and the axis of symmetry. The calculations indicate that the concept of the "collision angle" is ambiguous in the case of real shells that possess nonzero strength.

The average value of the angle (the angle between the line 1 and the axis of symmetry in Figure 5.14a) differs significantly from the angle of the slope of the stream in the immediate vicinity of the deceleration zone (the angle between the axis of symmetry and line 2 in Figure 5.14a). In this case, the first of these angles almost immediately reaches a stationary

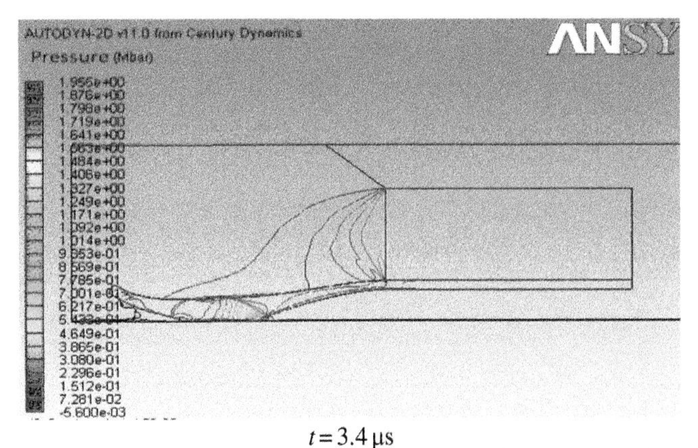

$t = 3.4\,\mu s$

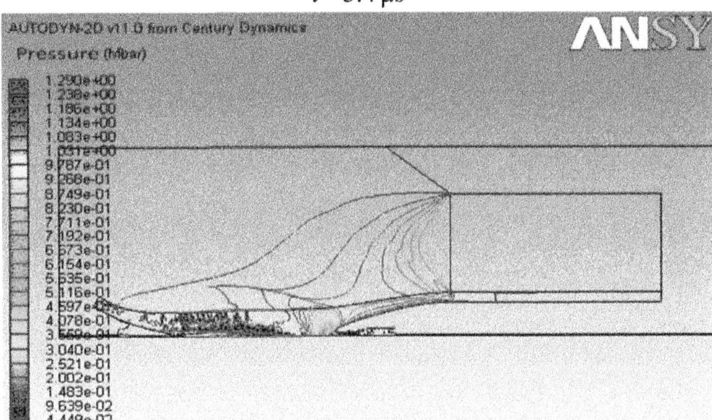

$t = 4\,\mu s$

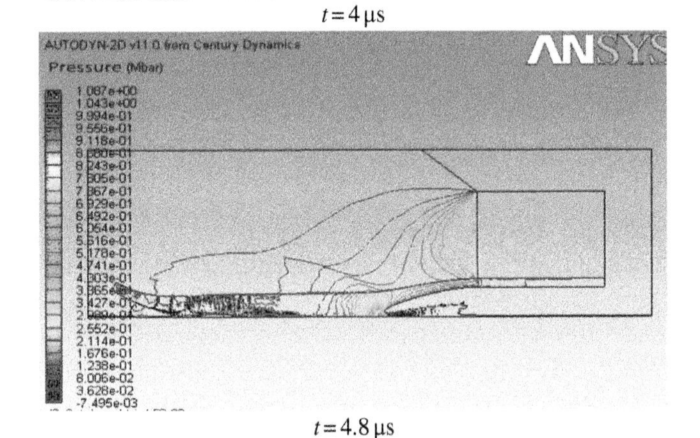

$t = 4.8\,\mu s$

Figure 5.13 Formation of a cumulative jet.

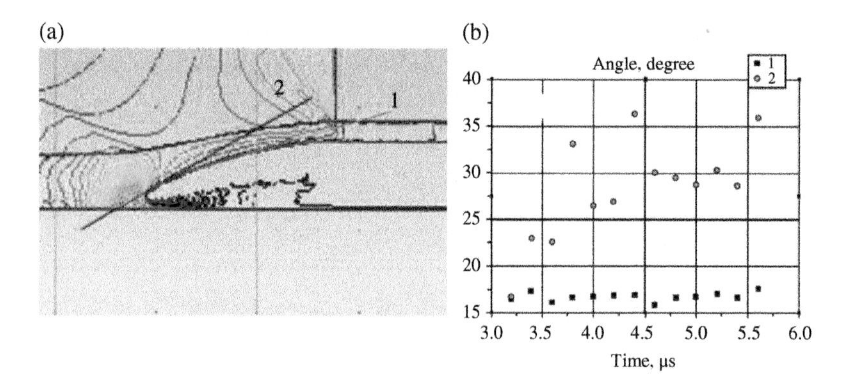

Figure 5.14 Changes of the collision angles.

value of approximately 16°, while the second one approaches a certain constant value of approximately 30° for a fairly long time, experiencing sharp jumps (Figure 5.14b). The latter confirms the nonstationary nature of the process in the vicinity of the flow deceleration zone.

The simulations which were closer to the experimental schemes were performed using the LEGAK method [15] and the kinetics of the detonation process developed by V.G. Morozov and I.I. Karpenko (MK-kinetics) [16]. In this case, the complete setup shown in Figure 5.1 was used for simulation of a ceramic tube collapse. The simulation process included the acceleration of the conical impactor, which initiated the explosive charge in a similar manner as in the experiment. Sequential stages of the process are shown in Figure 5.15. An interesting feature of the process is the formation of a cumulative jet from the material of the initiating impactor. Due to the sufficiently large angle of the cone, the effect is manifested in the form of "reverse cumulation" (see the first frame in Figure 5.15).

The resulting jet moves ahead of the elements of the converging tube, and for some time its elements are in the head part of the ceramic jet. As in the previous calculation, high-speed and main parts can be identified in the jet, and the jet is a stream of ceramics particles. The calculation shows the presence of a shock wave in the air produced by the motion of the jet. A comparison of the calculation results that take into account the kinetics of the detonation transformation with the calculations in which the forced initiation of steady-state detonation of explosives was introduced shows that the consideration of kinetics does not significantly affect the results of the simulation. When solving the problem of initiation of explosives by ceramic jets, this allowed us to use simpler calculation methods for the detonation of a charge acting on a tube. The results

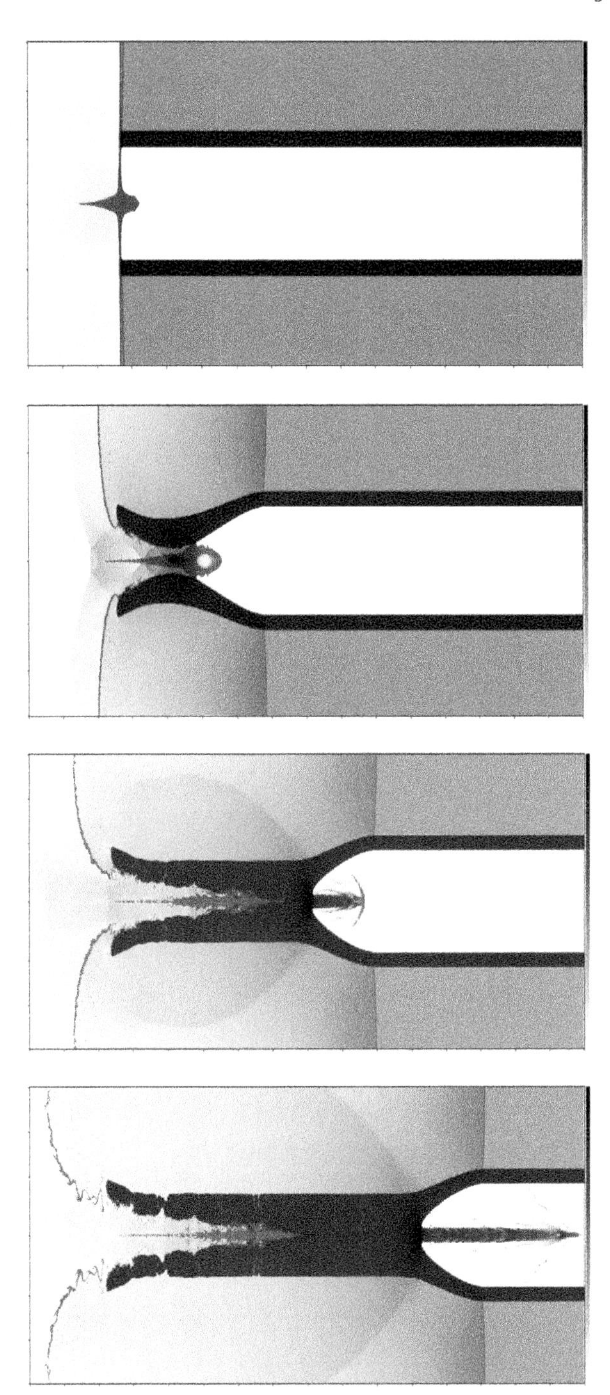

Figure 5.15 Formation of a cumulative jet.

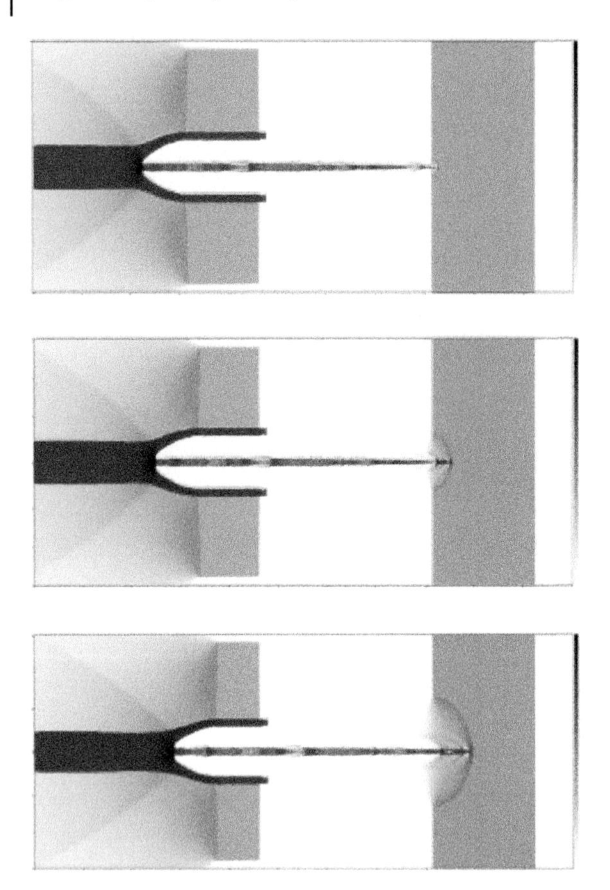

Figure 5.16 Initiation of the HE charge by the cumulative jet.

of solving the problem in the formulation of Figure 5.1 with an HMX charge (here the MK kinetics was used in calculation), located at 35 mm from the rear end of the charge with the tube, are shown in Figure 5.16. The first frame corresponds to the moment of the onset of the jet action on the charge. The following frames demonstrate the stages of the detonation development with the continued penetration of the jet (the zones behind the shock wave in the charge of the HMX are detonation products). Thus, the calculation confirms the experimental data.

It may be noted that a necessary condition for the formation of a cumulative jet in the explosive compression of ceramic tubes is the use of sufficiently powerful explosives with detonation pressure much higher than the strength properties of the ceramics.

5.3 The Effect of Hypervelocity Jet Impact Against a Steel Target

In the next series of experiments, the effect of ceramic jets formed during explosive compression of corundum tubes on steel targets was investigated [18]. A special method was proposed to produce ceramic tubes from corundum powder. The tubes were manufactured at the Institute of Hydrodynamics of the Siberian Branch of the Russian Academy of Science by detonation spraying using CCDS2000 equipment [19].

The electro-corundum powder having a particle size in range 20–28 μm was used for spraying. In the process of spraying, the powder particles were heated to a semi-molten state and accelerated toward the substrate at a velocity of 600–700 m/s [20]. During the process of interaction of individual particles between each other and with the substrate, a dense monolithic corundum layer was formed with a bonding strength of about 50 MPa. In one pass, a layer of about 10 μm was formed on the substrate and the total thickness was increased by spiral spraying. Copper tubes were used as the substrate. After the spraying, the substrate was subsequently dissolved in a solution of iron chloride. A copper tube covered with a layer of corundum is shown in Figure 5.17, the dimensions of corundum tubes are given in Figure 5.18.

Hydrostatic weighing showed that the porosity of the ceramic tubes was approximately 4%. The inner part of the tubes was impregnated with iron chloride, which was used to dissolve the copper. The influence of porosity and residual iron chloride on the results of the experiments was

Figure 5.17 Copper tube Ø15 × 1 mm, length −100 mm, corundum coating with a thickness of 1 mm. (*Source:* From Balaganskii et al. [18]. Reprinted with permission of Pleiades Publishing, Ltd.)

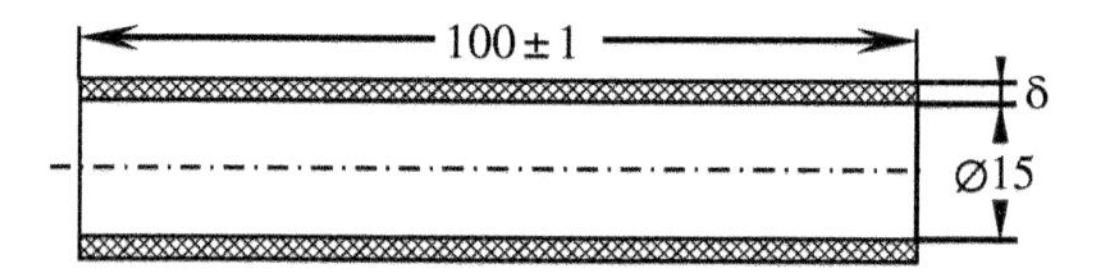

Figure 5.18 Dimensions of corundum tubes: $\delta = 0.5$, 1.0, and 1.5 mm. (*Source:* From Balaganskii et al. [18]. Reprinted with permission of Pleiades Publishing, Ltd.)

not evaluated. The internal diameter of all tubes was equal to the outer diameter of the copper tubes and was 15 mm, while the outer diameter depended on the thickness of the tube, which was $\delta = 0.5$, 1.0, or 1.5 mm.

The experimental setups were filled with explosive TG-50 by casting. The weight of each explosive charge was 200 g. The outer diameter of the charge was 36 mm, and the length of the charge was 100 mm. When conducting the experiments with 1.5 mm tubes, the charges had a shell of D16T alloy (Al-Cu-Mg) with a thickness of 5 mm. The scheme of the experimental assembly is shown in Figure 5.19 (the outer shell is not shown).

For comparison, an experiment was performed with a charge in a D16T shell with an axial channel, but without a tube. In this case, the action of the gas jet was evaluated. The charges were set at a distance of 50 mm from the target. As a target, disks of U7A steel with a diameter of 100 mm and a thickness of 19 mm were used. All experiments were performed at the Institute of Hydrodynamics of the Siberian Branch of the Russian Academy of Science. Figure 5.20 shows a photograph of all used tubes. Figures 5.21 and 5.22 show photographs of steel targets subjected to exposure of a cumulative jet.

A typical feature of the targets affected by the flow of ceramic particles was the unexpectedly large dimensions (diameter and depth) of the cavities.

The diameter of the cavities formed on the front side of the samples was 20–25 mm. The depth of the cavities increases with the thickness of the ceramic tube. When a charge with a 1.5 mm ceramic tube was used, a complete penetration of the sample was observed and there was also a cavity at the rear end of the sample. The total penetration depth was 30 mm. The target itself was destroyed due to the formation of radial cracks.

In the experiment with a hollow charge without the ceramic tube, a gas jet impacted the target. In this case, the defects of the sample were substantially smaller (Figure 5.23).

A sample deformed by a jet formed during explosive collapsing of a 0.5 mm tube (Figures 5.21-1 and 5.22-1) was investigated by optical and electron microscopy.

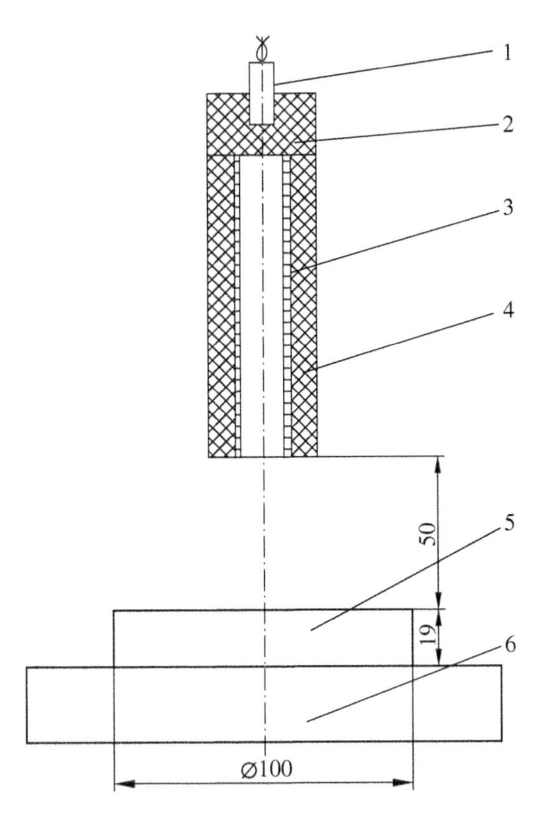

Figure 5.19 The scheme of the experimental assembly: 1, electric detonator; 2, initiating charge; 3, main HE charge; 4, ceramic tube; 5, target; and 6, base. (*Source:* From Balaganskii et al. [18]. Reprinted with permission of Pleiades Publishing, Ltd.)

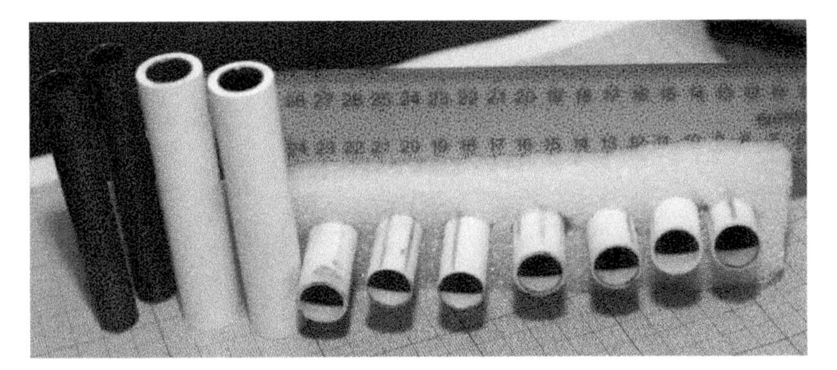

Figure 5.20 Tubes used in the experiments. (*Source:* From Balaganskii et al. [18]. Reprinted with permission of Pleiades Publishing, Ltd.)

Figure 5.21 Face sides of the targets subjected to the action of jets formed from ceramic tubes of different thicknesses: 1, $\delta = 0.5$ mm; 2, $\delta = 1.0$ mm; and 3, $\delta = 1.5$ mm. (*Source:* From Balaganskii et al. [18]. Reprinted with permission of Pleiades Publishing, Ltd.)

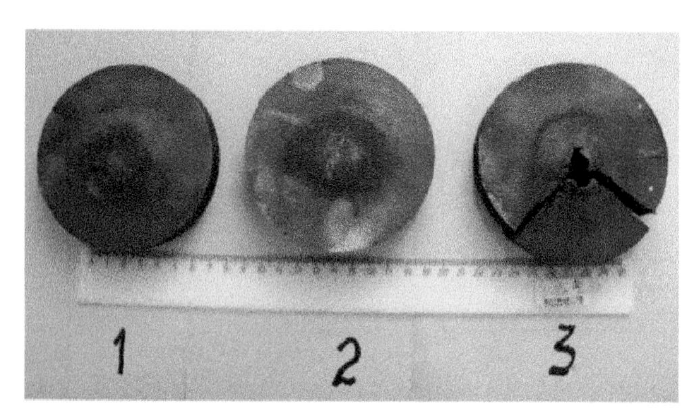

Figure 5.22 The back sides of the targets subjected to the action of jets formed from ceramic tubes of different thicknesses: 1, $\delta = 0.5$ mm; 2, $\delta = 1.0$ mm; and 3, $\delta = 1.5$ mm. (*Source:* From Balaganskii et al. [18]. Reprinted with permission of Pleiades Publishing, Ltd.)

Figure 5.23 The face (left) and back (right) side of the target subjected to loading with a gas-cumulative jet. (*Source:* From Balaganskii et al. [18]. Reprinted with permission of Pleiades Publishing, Ltd.)

The detailed analyze the structural transformations due to impact loading is given in Section 7.4.

The formation of spherical pores indicates that the material reached the boiling point. It is known that the boiling point of pure iron is about 2750 °C, but when the alloying elements typical for steel are added, the boiling point of the alloy can decrease. Nevertheless, the presented data indicate that the growth of cracks formed on the surface of the cavity occurred with an extremely high velocity under conditions close to adiabatic. Subsequent heat removal into the adjacent areas led to quenching and formation of a martensitic–austenitic structure along the cracks.

To explain the results of the experiments on the effect of ceramic jets on steel targets, numerical simulation of the explosive compression of corundum tubes with different wall thicknesses was performed. The formulation of the problems corresponded to Figure 5.19. The geometric parameters of computational domains fully corresponded to the experimental setups. The simulation was performed using ANSYS AUTODYN software [21].

Having a goal to correctly reproduce the physical processes occurring during the explosive collapse of the ceramic tube and the interaction of the cumulative jet with the target, at the first stage of simulation, the EOS parameters were verified. This was performed by comparing the simulation results with the experimental data described in [9, 10]. Based on the comparative analysis, those parameters of the governing relationships were chosen, which provided a good agreement between calculated and measured velocities of the jet and the rate of penetration of the jet into the water. The parameters of the shock adiabat of corundum for pressures up to 300 GPa were found in the handbook by R.F. Trunin et al. [22].

The simulation was conducted in a two-dimensional axisymmetric formulation on a stationary Eulerian mesh. The lower boundary of the calculated area was the symmetry axis. The spatial resolution was 20 cells/mm. The detonation was initiated from the left side of the explosive charge. The models used in the simulation in terms of AUTODYN software are presented in Table 5.1. The parameters for corundum, which are necessary for modeling the explosive collapse process, are given in Table 5.2.

Table 5.1 Relations used for the calculation of all involved substances.

Material name	Equation of state	Strength	Failure
Composition B	JWL	None	None
Al_2O_3	Shock	von Mises	Hydro (Pmin)
Steel 1006	Shock	Johnson Cook	None
Al 2024	Shock	None	None

Source: From Balaganskii et al. [18]. Reprinted with permission of Pleiades Publishing, Ltd.

Table 5.2 Constants describing the behavior of corundum.

EOS	Shock
Reference density	3.92 g/cm^3
Gruneisen coefficient	0.5
Parameter C_1	0.871 cm/μs
Parameter S_1	0.713
Strength	**von Mises**
Shear modulus	1.0 Mbar
Yield stress	0.08 Mbar
Failure	**Hydro P_{min}**
P_{min}	−0.0075 Mbar

Source: From Balaganskii et al. [18]. Reprinted with permission of Pleiades Publishing, Ltd.

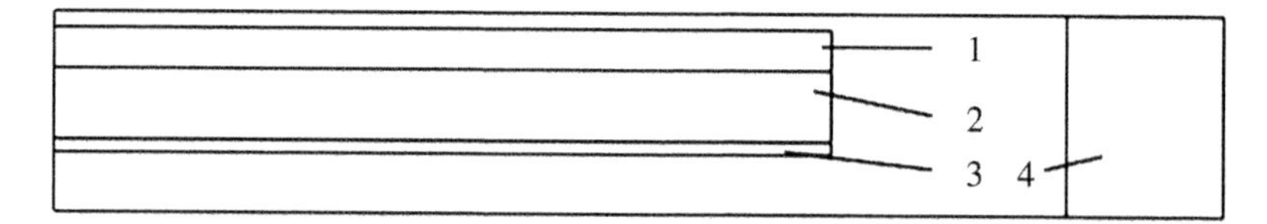

Figure 5.24 Initial posting of the problem: 1, duralumin shell; 2, HE charge; 3, ceramic tube; and 4, steel target. (*Source:* From Balaganskii et al. [18]. Reprinted with permission of Pleiades Publishing, Ltd.)

At the stationary stage of the jet formation, all three variants of the simulation with tubes of different thickness are characterized by similar pressures at the collision point and the velocity of the cumulative jet. Consequently, the simulation results are discussed in detail below by considering only the setup with a 1.5 mm ceramic tube.

The geometry of the simulation domain at the initial state is shown in Figure 5.24. It should be noted that explosive charge is surrounded by a shell. Figures 5.25–5.27 show the fields of material flows and pressure contours at specific moments of time.

The pressure at the point of contact at the stationary stage is approximately 0.9–1.0 Mbar, the maximum velocity of the leading part of the jet is approximately 23 km/s, and the velocity of the main jet is 14 km/s. The calculated temperature at the penetration zone exceeds 31 000 K.

In all analyzed cases, the collision parameters at the moment of the interaction of the jet with the target go far beyond the applicability of the EOS of the steel target. This leads to anomalous values of the pressure

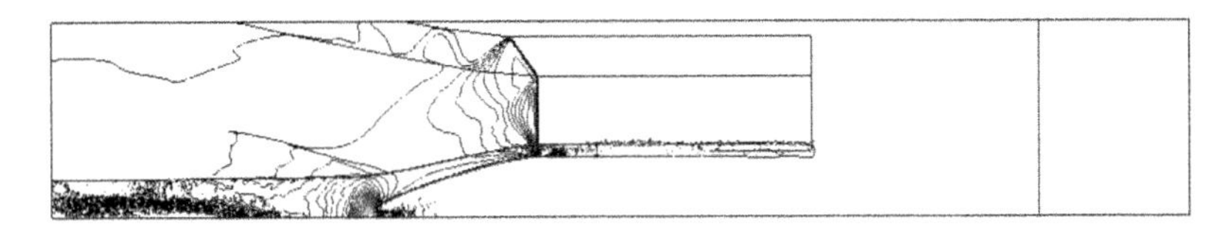

Figure 5.25 Flow fields and pressure contours at the stationary stage at time $t = 8\,\mu s$. (*Source:* From Balaganskii et al. [18]. Reprinted with permission of Pleiades Publishing, Ltd.)

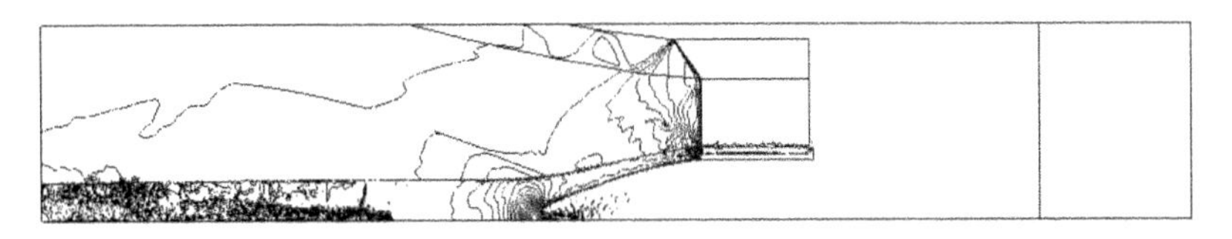

Figure 5.26 Flow fields and pressure contours at time $t = 10.25\,\mu s$. (*Source:* From Balaganskii et al. [18]. Reprinted with permission of Pleiades Publishing, Ltd.)

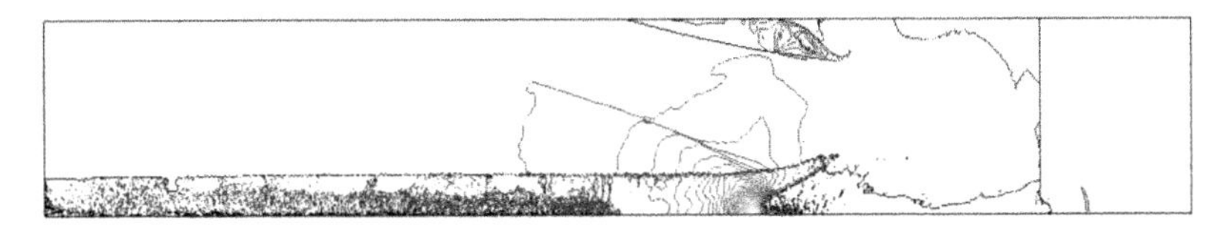

Figure 5.27 The beginning of the penetration of the jet into the target at time $t = 14.25\,\mu s$. (*Source:* From Balaganskii et al. [18]. Reprinted with permission of Pleiades Publishing, Ltd.)

and temperature in the collision zone. Apparently, to explain the results obtained, it is necessary to use a wide range EOS, which takes into account phase transitions such as melting and evaporation in a rarefaction wave.

5.4 Modeling of Fast Jet Formation Under Explosion Collision of Two-Layer Alumina/Copper Tubes

In our opinion, the explosive collision of two-layer tubes with an outer layer of high-modulus ceramics and an inner layer of copper can lead to the formation of high-speed continuous copper jets [23, 24]. Numerical simulation using ANSYS AUTODYN was performed in a 2D axisymmetric formulation on a Eulerian mesh of 3900 × 1200 cells. The spatial resolution was 30 cells/mm. The detonation wave was initiated as a plane wave and propagated from the left to the right boundary of the mesh. The boundary conditions on the left, right, and upper bounds were

Table 5.3 Relations used for the calculation of all used substances.

Material name	Equation of state	Strength	Failure
AL 921-T	Shock	None	None
COMP B 163	JWL	None	None
Copper1	Shock	Johnson Cook	None
AL2O3 CERA	Shock	von Mises	None
STEEL 1006	Shock	Johnson Cook	None

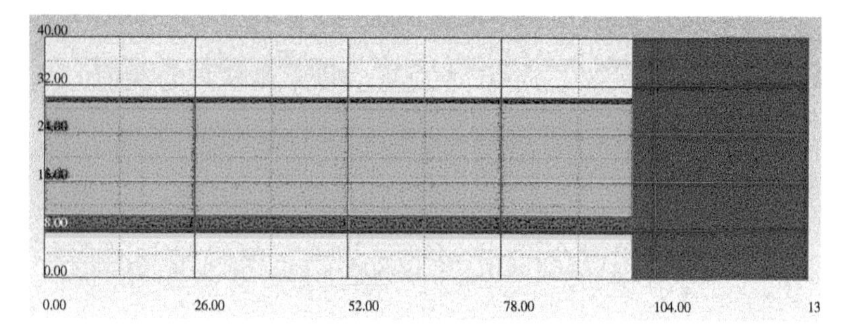

Figure 5.28 Initial posting with dimensions of 130 × 40 mm. The aluminum shell is shown in dark blue; HE charge – green; ceramic tube – red; copper tube – light blue; steel target – purple.

defined in the AUTODYN terminology as "flow out." The models for materials were chosen from the AUTODYN EOS library and are given in Table 5.3.

Initial posting with dimensions of 130 × 40 mm is shown in Figure 5.28. The detonation was initiated on the left surface of the explosive charge. The results of the simulation are given in the form of isolines, flow fields, and spatial and temporal graphs of the main parameters.

Figure 5.29 shows the flow fields and pressure contours at the time of the tube collapsing to the symmetry axis at $t = 7.6\,\mu s$. Figure 5.30 shows the distribution of radial velocity in the thickness of the outer shell in the range between 0 and 10 mm at $t = 7.6\,\mu s$. Figure 5.31 shows the distribution of pressure across the thickness of the outer shell from 0 to 10 mm at $t = 7.6\,\mu s$.

It is easy to observe that the tube collapses at the symmetry axis at a speed of about 2 km/s and the pressure exceeds 700 GPa.

We should particularly note that the calculated temperature at the point of contact at the initial stage of the collision exceeds 20 000 K.

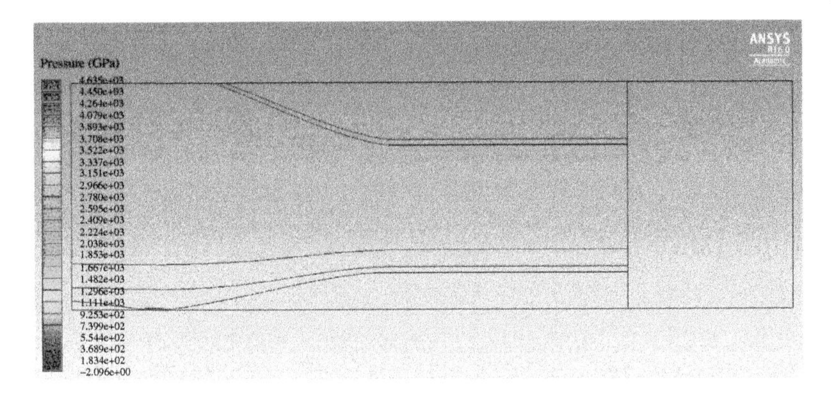

Figure 5.29 Material flow fields and pressure contours at the time of the collapse of the tube to symmetry axis at $t = 7.6\,\mu s$.

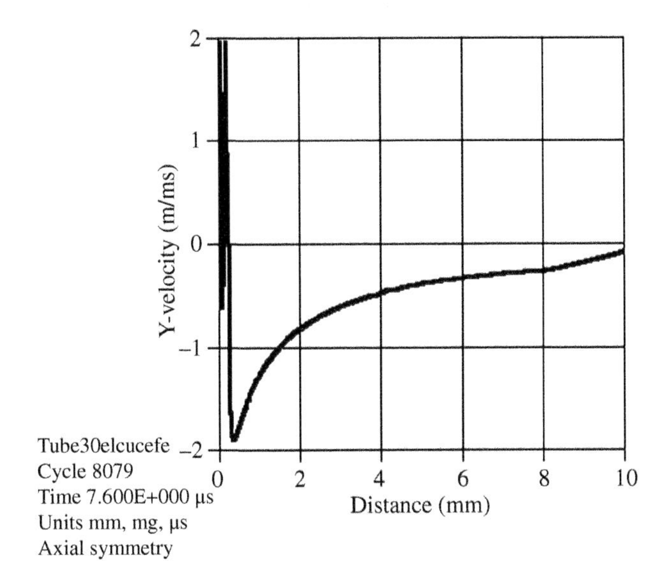

Figure 5.30 Radial velocity profiles versus tube thickness from 0 to 10 mm at $t = 7.6\,\mu s$.

Fields of material flows and pressure profiles at $t = 8.5\,\mu s$ are given in Figure 5.32. The particle velocity and pressure profiles along the symmetry axis from 25 to 50 mm are shown in Figures 5.33 and 5.34, respectively.

The velocity along the length of the jet varies from 21.3 to 28.9 mm/μs. Maximal pressure reaches 540 GPa. The process of collapse of the tube is of a nonstationary nature.

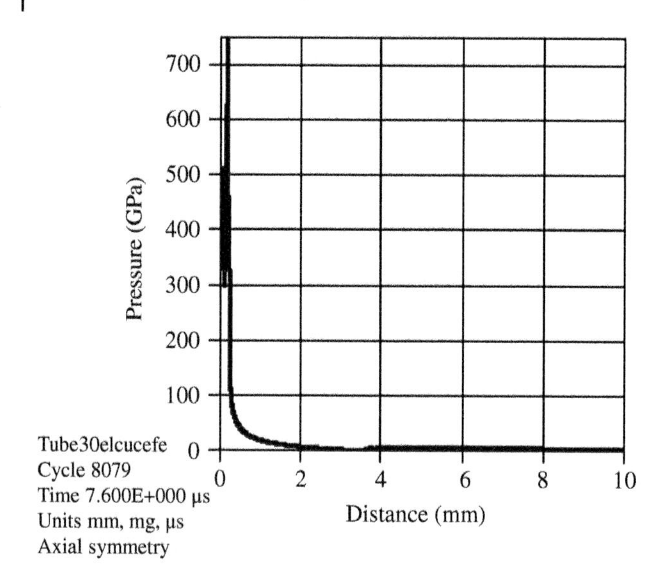

Figure 5.31 Pressure profiles versus tube thickness from 0 to 10 mm at $t = 7.6\,\mu s$.

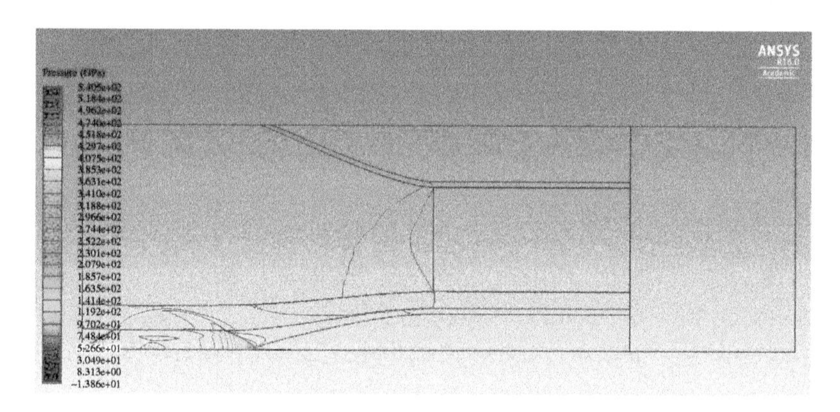

Figure 5.32 Material flow fields and pressure contours at $t = 8.5\,\mu s$.

Figure 5.35 shows the flow fields and pressure contours at the time when the jet penetrates the target at $t = 10.7\,\mu s$. Figures 5.36 and 5.37 show the velocity and pressure distributions along the length of the jet. Figures 5.38 and 5.39 give the distribution of temperature and effective plastic deformation along the length of the jet at the same moment of time.

The leading part of the jet, formed at the nonstationary stage, is broken and consists of separate fragments. The leading part of the dense stream

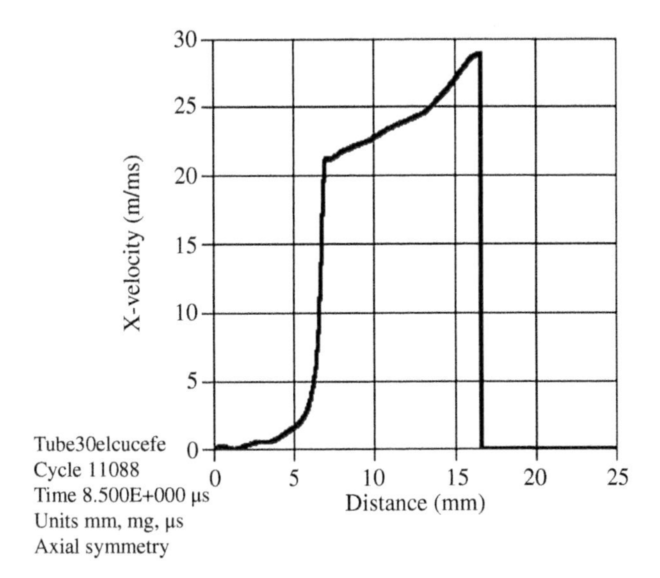

Figure 5.33 Particle velocity profile along the symmetry axis from 25 to 50 mm at $t = 8.5\,\mu s$.

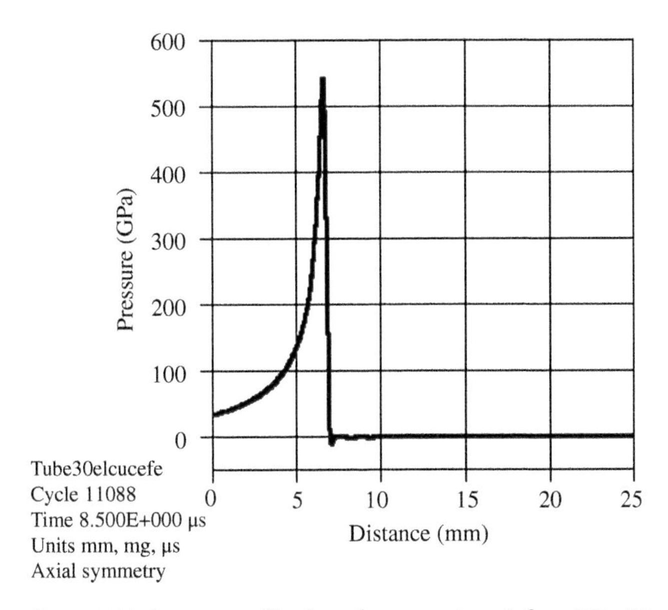

Figure 5.34 Pressure profile along the symmetry axis from 25 to 50 mm at $t = 8.5\,\mu s$.

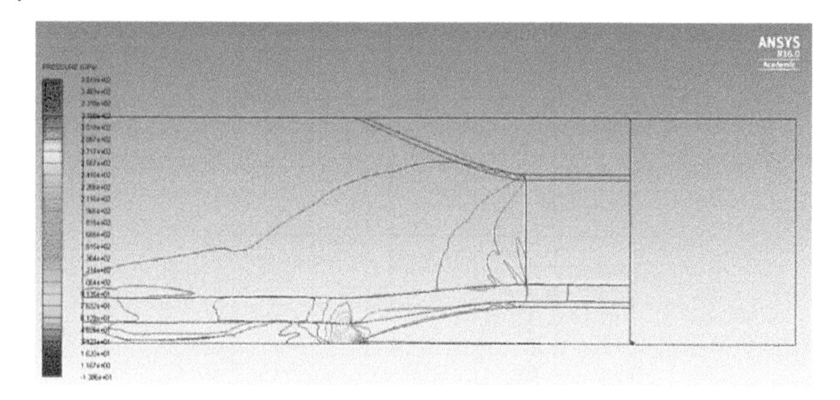

Figure 5.35 Material flow fields and pressure contours at the beginning of the process of jet penetration into target $t = 10.7\,\mu s$.

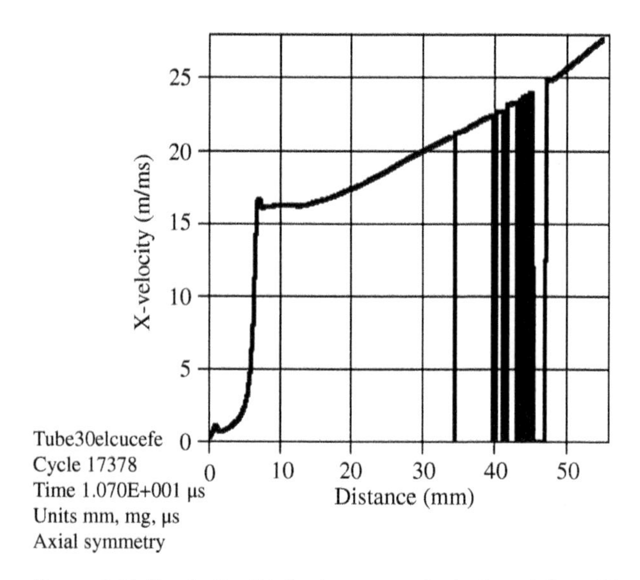

Figure 5.36 X-velocity distribution versus the jet length from 50 to 100 mm at $t = 10.7\,\mu s$.

moves at a speed of about 20 mm/μs. The temperature of this part is about 2000 K, the jet tears apart when the effective plastic deformation reaches 30.

Figure 5.40 shows the flow fields and pressure contours at the time when the jet penetrates the target $t = 14.5\,\mu s$.

Figures 5.41–5.44 show the distribution of velocity, pressure, temperature, and effective plastic strain along the length of the jet from $x = 75$ to $x = 130$ mm at the same moment of time.

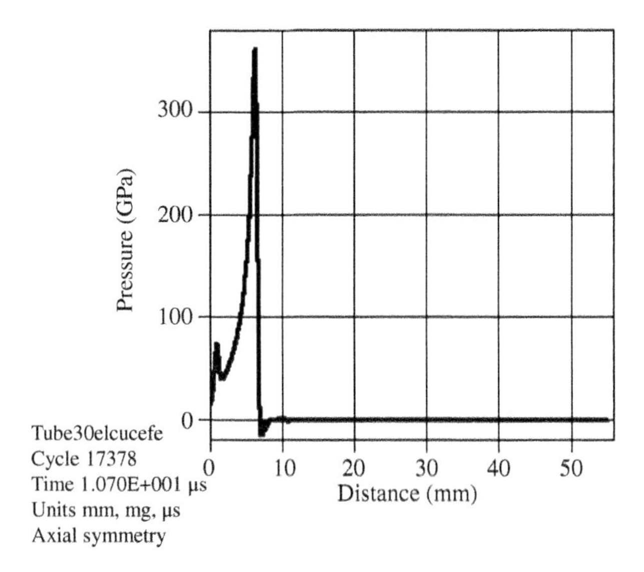

Figure 5.37 Pressure distribution versus the jet length from 50 to 100 mm at $t = 10.7\,\mu\text{s}$.

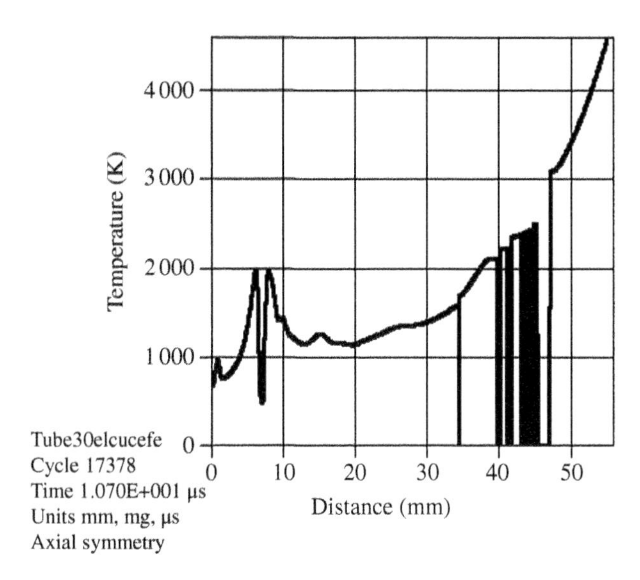

Figure 5.38 Temperature distribution along the length of the jet from 50 to 100 mm at $t = 10.7\,\mu\text{s}$.

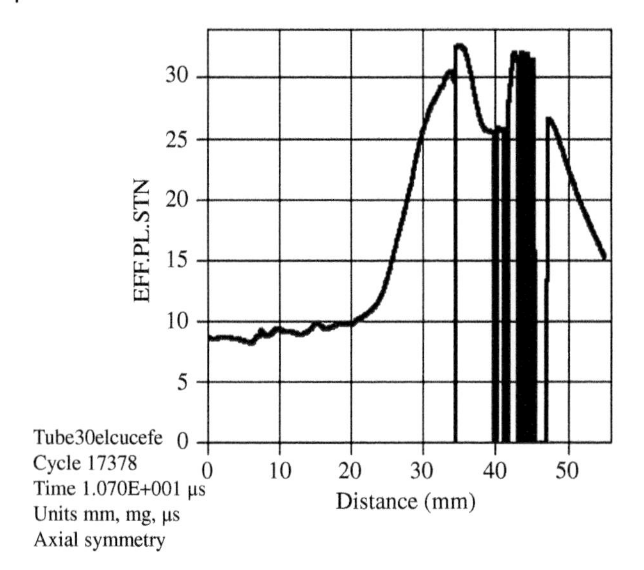

Figure 5.39 Effective plastic strain distribution along the length of the jet from 50 to 100 mm at $t = 10.7\,\mu s$.

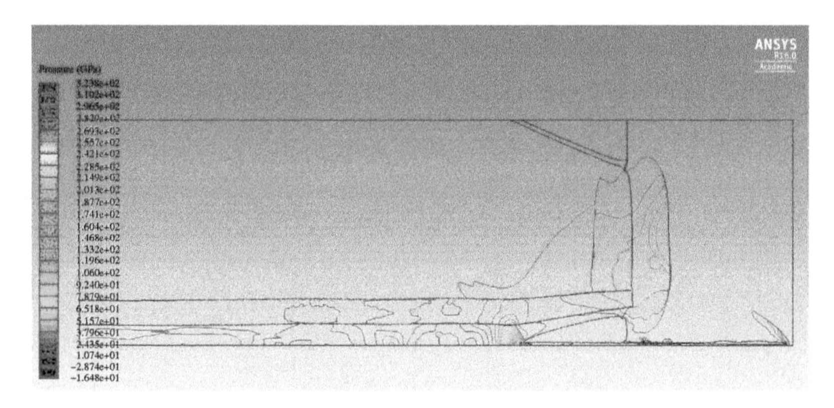

Figure 5.40 Flow fields and pressure contours at time $t = 14.5\,\mu s$.

5.5 Summary

The results of the first series of experiments with ceramic tubes showed that when the tube collapses, a brightly glowing particle stream (conventionally called a cumulative jet) is formed. The head part of the jet propagates at a speed about two times higher than the detonation velocity. The average density of ceramic particle flow is about $0.33\,g/cm^3$. Two

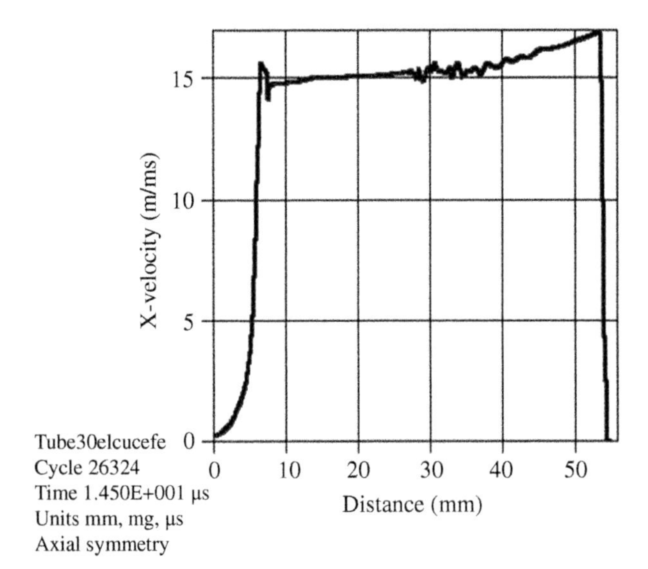

Figure 5.41 X-velocity distribution along the length of the jet from 75 to 130 mm at $t = 14.5\,\mu s$.

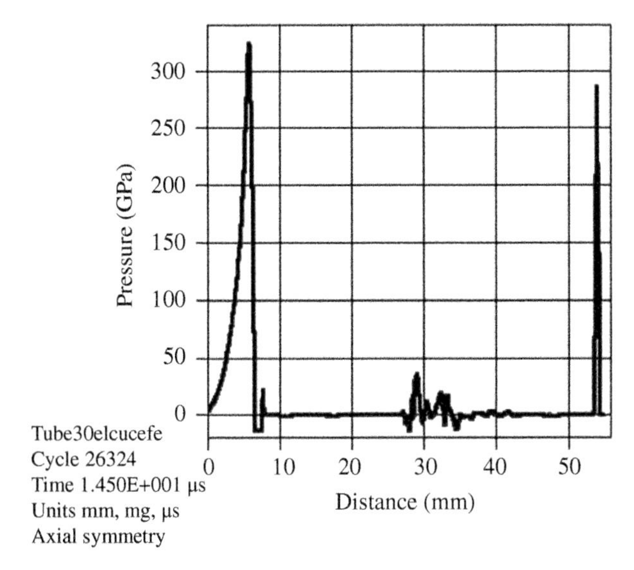

Figure 5.42 Pressure distribution along the length of the jet from 75 to 130 mm at $t = 14.5\,\mu s$.

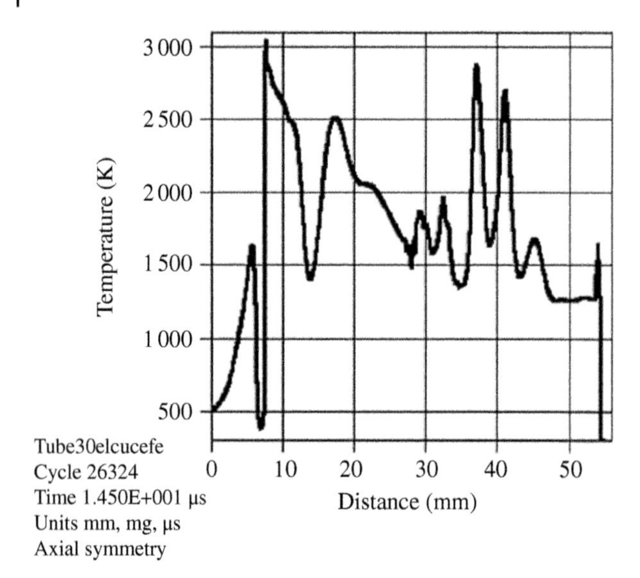

Figure 5.43 Temperature distribution along the length of the jet from 75 to 130 mm, $t = 14.5\ \mu s$.

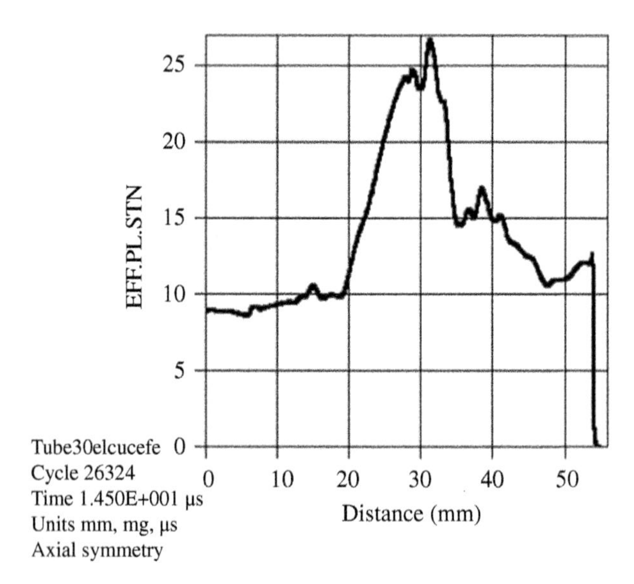

Figure 5.44 Effective plastic strain distribution along the length of the jet from 75 to 130 mm, $t = 14.5\ \mu s$.

glowing fronts moving with velocities of approximately 12.4 and approximately 11 km/s were recorded.

The formed stream (jet) of particles has an initiating ability. In the case of shock wave loading of high-modulus ceramic rings, a cumulative effect is also observed.

The numerical simulation also shows that when a ceramic tube collapses, a cumulative jet is formed, almost immediately breaking up into separate fragments, which corresponds to experimental observations. The head particles have a substantially higher velocity than the main mass of the jet, which also agrees with the experimental data. An important characteristic that significantly influences the formation of a cumulative jet is the collision angle. In the axisymmetric case, the angle between the generatrix of the collapsing shell and the axis of symmetry is of great importance. The calculations indicate the ambiguity of the "collision angle" concept in the case of real ceramic shells. The average value of the angle differs significantly from the inclination angle of the flow in the immediate vicinity of the deceleration zone. In the same time, the first of these angles almost immediately reaches the constant value of approximately 16°, while the second approaches close enough to the value of approximately 30°, experiencing sharp jumps, which indicates that the process is unsteady in the vicinity of the deceleration zone of the flow. The calculation confirms that the particle flow has an initiating power and demonstrates the stages in the development of the detonation process as the jet enters the charge.

In the second series of experiments, an estimation of the penetrating effect of a discrete ceramic jet produced by the explosive compression of ceramic tubes was made. A specific feature of the targets affected by the flow of ceramic particles was the anomalously large dimensions (diameter and depth) of the emerging caverns.

The melting and boiling of the target material were observed using structural analysis methods. The boiling of the target material indicates that the collision velocity of the jet with the target exceeded some threshold velocity leading to "thermal explosion."

The results of the numerical simulation are consistent with the results of experimental observation and material characterization. The maximum velocity of the leading part of the jet is approximately 23 km/s, and the speed of the main jet is 14 km/s. The results of simulation indirectly confirm the experimental data, which testify to extremely high temperatures and pressures in the region the jet interacts with the target.

Based on the simulation results of the two-layer tube compression, two stages of the collapse process were identified: the nonstationary and stationary stages. At the initial stage, an unsteady fragmented jet is formed with the velocity of the leading elements up to 30 km/s. The velocity of the collapse of the tube on the axis of symmetry is about 2 km/s, the pressure in the contact zone exceeds 700 GPa. We should particularly note that the calculated temperature at the point of contact at the initial stage of the collision exceeds 20 000 K.

At the subsequent stationary stage, a continuous jet is formed having a speed of about 20 km/s. The temperature of this continuous jet is about 2000 K. The jet breaks up when the value of effective plastic strain reaches 30.

References

1 Koski, W.S., Lucy, F.A., Shreffler, R.G. et al. (1952). Fast jets from collapsing cylinders. *Journal of Applied Physics* 23: 1300–1305.
2 Kinelovskii, S.A. and Trishin, Y.A. (1980). Physical aspects of the hollow-charge effect. *Combustion, Explosion and Shock Waves* 16: 504–515.
3 Walters, W.P. and Zukas, J.A. (1989). *Fundamentals of Shaped Charges*. New York: Willey.
4 Titov, V.M., Fadeenko, Y.I., and Titova, N.S. (1968). Acceleration of solid particles by cumulative explosion. *Doklady Akademii Nauk SSSR* 180 (5): 1051–1052. [in Russian].
5 Shell, R. (1967). Detonation physics. In: *Kurzzeitphysik*, vol. 2 (ed. K. Vollrath and G. Thomer), 276–349. Wien: Springer.
6 Trishin, Y.A. (2000). Some physical problems of cumulation. *Journal of Applied Mechanics and Technical Physics* 41: 773–787.
7 Trishin, Y.A. (2005). *Physics of Cumulative Processes*. Novosibirsk: Institute of Hydrodynamics SB RAS [in Russian].
8 Balagansky, I.A., Agureikin, V.A., Nosenko, N.I. et al. (1998). Accelerating device based on high explosive charge which contains high modulus ceramic tube. *Collected Scientific Works of the NSTU*, vol. 2, 27–32. [in Russian].
9 Balagansky, I.A., Agureikin, V.A., Kobilkin, I.F. et al. (1999). Acceleration device based on high explosive charge, which contains high modular ceramic tube. *International Journal of Impact Engineering* 22: 813–823.
10 Balagansky, I.A., Vinogradov, A.V., Agureikin, V.A. et al. (2002). Explosion systems based on HE explosion charges, which include high-modulus ceramic tubes. *Proceedings of the International conference 'III Khariton Thematic Scientific Readings'*, Sarov, Russia (26 February–2 March 2001). Sarov: RFNC. [in Russian].
11 Balagansky, I.A., Vinogradov, A.V., Kobylkin, I.F. et al. (2006). Formation of the cumulative particle flow under shock-wave loading of a high-modulus ceramics ring. *Proceedings of the 7th All-Russian Scientific and*

Technical Conference 'Science. Industry. Defense', Novosibirsk, Russia (19–21 April 2006). Novosibirsk: Novosibirsk State Technical University. [in Russian].

12 Vinogradov, A.V., Merzhievsky, L.A., and Balagansky, I.A. (2010). Modeling of cumulative processes under the collapse of ceramic liners. *Abstracts of International conference 'Zababakhin scientific readings'*, Snezhinsk, Russia (15–19 March 2010). Snezhinsk: RFNC. [in Russian].

13 Vinogradov, A.V., Merzhievsky, L.A., and Balagansky, I.A. (2012). Modeling of compression of a ceramic tube by detonation products. *Dynamics of a Continuous Medium* 127: 21–24. [in Russian].

14 Merzhievsky, L.A., Balagansky, I.A., Vinogradov, A.V. et al. (2011). Cumulative effects under compression of a ceramic tube by detonation products. *Proceedings of the international conference 'XIII Khariton Thematic Scientific Readings'*. Sarov, Russia (14–18 March 2011). Sarov: RFNC. [in Russian].

15 Bakhrakh, S.M., Velichko, S.V., Spiridonov V.F. et al. (2004). Methodology LEGAK-3D for calculation of three-dimensional non-stationary flows of multicomponent continuous medium, and the principles of its implementation in multiprocessor computers with distributed memory. *VANT. Series 'Mathematical Modeling of Physical Processes'* 4: 41–50. [in Russian].

16 Morozov, I.I., Karpenko, S.E., Kuratov, S.S. et al. (1995). The theoretical justification of the phenomenological model of shock wave sensitivity of heterogeneous explosives based on TATB taking into account the one- and two-times the shock-wave loading. *Chemical Physics* 14 (1–2): 32–39. [in Russian].

17 Agureikin, V.A. and Kryukov, B.P. (1986). The method of individual particles for the calculation of flows of multicomponent media with large deformations. *Numerical methods of continuum mechanics.* 17 (4): 17–31. [in Russian].

18 Balaganskii, I.A., Merzhievskii, L.A., Ul'yanitskii, V.Y. et al. (2018). Generation of hypervelocity particle flows by explosive compression of ceramic tubes. *Combustion, Explosion and Shock Waves* 54: 119–124.

19 Shtertser, A., Ulianitsky, V., Zlobin, S. et al. Computer controlled detonation spraying of WC/Co coatings containing MoS_2 solid lubricant. *Surface & Coatings Technology* 206: 4763–4770.

20 Ulianitsky, V., Shtertser, A., Sadykov, V. et al. (2016). Development of catalytic converters using detonation spraying. *Materials and Manufacturing Processes* 31 (11): 1433–1438.

21 Century Dynamics (2005). *Autodyn. Explicit Software for Nonlinear Dynamics.* Theory manual.

22 Trunin, R.F., Gudarenko, L.F., Zhernokletov, M.V. et al. (2001). *Experimental DATA on Shock Compression and Adiabatic Expansion of Condensed Matter*. Sarov: RFNC-VNIIEF.

23 Balagansky, I.A. Merzhievsky, L.A., and Vinogradov, A.V. (2016). Modelling of fast jet formation under explosion collision of two-layers alumina/copper tubes. *Proceedings of Fifth international symposium on explosion, shock wave and high-strain-rate phenomena*. Beijing, China (25–28 September 2016). Beijing: Beijing Institute of Technology.

24 Balagansky, I.A., Vinogradov, A.V., and Merzhievsky, L.A. (2017). Modelling of fast jet formation under explosion collision of two-layer alumina/copper tubes. *The International Journal of Multiphysics* 11 (3): 265–275.

6
Protective Structures Based on Ceramic Materials

One of the relevant problems for explosive technologies is designing and manufacturing containers and shells protecting dangerous substances from unauthorized external influences during their transportation and storage, such as a fragmentation bullet impact, the action of shaped charges, shock waves, etc. A variety of protective shells was proposed that are mainly based on heterogeneous materials, such as highly porous refractory ceramics, various foams, air suspension of dispersed media, fine granular materials, etc. They were selected for their good energy-absorbing properties under the influence of dynamic loads [1–4].

Recently, the demand for the high energy-absorbing capacity of materials used in protective shells and containers has been supplemented by increased requirements for their heat, chemical, and radiation resistance [5]. Powders of ceramic materials, possessing high deformation characteristics, meet these additional requirements to a substantial degree. ZrO_2, MgO (chamotte), and Al_2O_3 (corundum) powders [1] and aluminosilicate microspheres [3] were considered for the application as mechanical and thermal load dampeners. A logical continuation of those studies was the study of the possibility to create protective structures based on ceramic powders.

6.1 Detonation Transmission Through Dispersed Ceramic Media

One of the most important properties of a shell material that protects explosives from accidental detonation is its ability to transmit detonation. The phenomenon of detonation initiation in an explosive charge located at some distance from the detonating charge was discovered in

Explosion Systems with Inert High-Modulus Components: Increasing the Efficiency of Blast Technologies and Their Applications, First Edition. Igor A. Balagansky, Anatoliy A. Bataev, and Ivan A. Bataev.

the middle of the last century and was named "transfer of detonation by influence" [6]. In this case, the active (initiating) charge can be separated from the passive one (in which the detonation is initiated) by both an air gap and a barrier of dense inert material. The mechanism of detonation transfer is qualitatively similar for various transfer media and types of explosives for passive charges. The detonation is initiated by shock waves propagating through the inert barrier.

Studies of detonation transmission through air gaps or dense media have both scientific and practical aspects [7].

From the scientific point of view, this phenomenon is used to study the initiation of explosives by shock waves. In this case, the main goal is to obtain the sensitivity characteristics of explosives for shock wave action (initiation criteria), which are subsequently used for safety assessments and for applications in detonation models. These results help to understand the kinetics of the detonation transformation. For practical purposes, an important task is finding the distances that are safe with regards to detonation transmission and evaluation of the parameters of barriers preventing explosives from accidental detonation during storage and transport to facilitate the development of explosion-proof containers [1–4, 8].

The experimental setup corresponds to the simplest but widely used method for studying the critical conditions for detonation initiation, which involves loading the investigated passive explosive charge with a contact explosion of an active charge separated from the passive one by an inert barrier or an air gap (Figure 6.1). In English-language literature, this method is usually called the GAP test; in the Russian literature, it is often referred to as the barrier method.

In the first series of experiments [10–13], the transmission of detonation through heterogeneous dispersive media based on silicon carbide

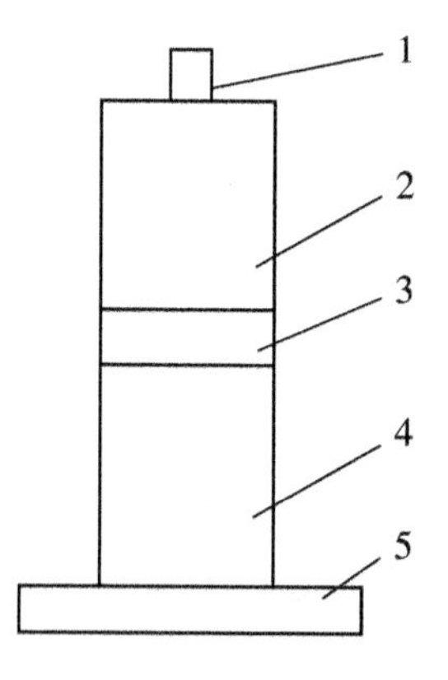

Figure 6.1 The scheme of the experimental assembly: 1, detonator; 2, active high explosive (HE) charge; 3, barrier of inert material; 4, charge of the investigated HE; and 5, witness plate. (*Source:* From Merzhievskii et al. [9]. Reprinted with permission of Pleiades Publishing, Ltd.)

(SiC) powder and river sand (silica, SiO_2) in a dry and water-filled state was studied. A photo of the powders showing the differences in their fractional composition is shown in Figure 6.2 (silica sand in the upper part). A paper clip made of a 0.8 mm steel wire is shown for scale.

The photo clearly shows that the silica particles are much larger. The dimensions of the sand particles lie in the range between 0.1 and 1.0 mm. Particles of silicon carbide have a more uniform granulometric composition; their sizes lie in the 0.05–0.1 mm range. The experimental assembly consisted of two cardboard cylindrical cups, one tightly inserted into the other. Between the bottoms of the cups, the powders were placed. The outer diameter of the assembly is 80 mm, and the thickness of the bottom is 0.75 mm. A general view of the assembly with a container and charges is shown in Figure 6.3.

After filling the powder, it was compacted with a 2 kg weight. It was found that the complete compaction under such conditions occurs approximately in 30 minutes. With an increase in the compression time to 24 hours, the thickness of the layer did not change. The densities of dry sand and silicon carbide powder after pre-compression were 1.85 and 1.87 g/cm^3, respectively. After compaction, water was poured into the assembly until it was completely saturated to produce water-filled buffers. The density of the wet powder was not measured. The data in [14] indicates that water-saturated sand contains about 20–30% of water. One experiment was also performed to transfer the detonation through a 12 mm thick silicon carbide plate placed between two cardboard pads similarly to the experiments with powders.

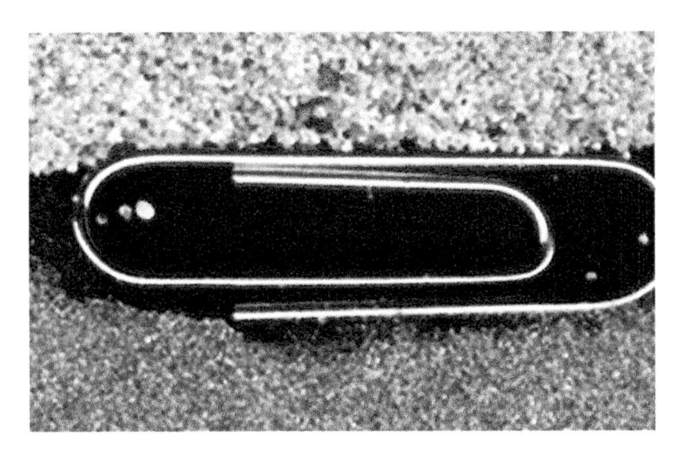

Figure 6.2 Fractional composition of powders. (*Source:* From Merzhievskii et al. [9]. Reprinted with permission of Pleiades Publishing, Ltd.)

Figure 6.3 Photograph of the experimental assembly. (*Source:* From Merzhievskii et al. [9]. Reprinted with permission of Pleiades Publishing, Ltd.)

The transmission of detonation for pairs (active/passive) of TNT/TNT and cast TG-40/TG-40 high explosive (HE) charges was studied. The height and diameter of the cylindrical charges were 40 mm, the mass of the TG-40 charge was 83 g, the density was 1.65 g/cm^3, the mass of the TNT charge was 79 g, and the density was 1.57 g/cm^3. The initiation or the absence of detonation in a passive charge was judged by the presence of an imprint from the detonation products on the "witness" plate. The results of the first series of experiments are shown in Table 6.1.

Here and further h_1 is the thickness of the inert layer, at which detonation is steadily initiated in a passive charge, h_2 is the thickness at which detonation can still occur, h_3 is the thickness at which detonation is absent. P_k is the calculated pressure value in the damping layer at the boundary with the active explosive charge. The calculation was performed according to the classical scheme of the disintegration of the discontinuity between the detonation products and the inert barrier. With the known equations of the state of the products and the shock adiabat (Rankine–Hugoniot curve) of the inert medium, this calculation does not pose any difficulties. Polytropic approximation was used for detonation products. Shock adiabats of dry and water-saturated sand are given in [14, 15]. For the barriers of silicon carbide powder using the methods of [16, 17], approximate shock adiabats were constructed. To verify the correctness of the method, the shock adiabats SiO_2 with several values of porosity and for SiC with a density of 2.3 g/cm^3 were constructed by the same method and compared with the experimental shock adiabats given

Table 6.1 Results of experiments of the first series.

High explosive (HE)	Material of barrier	State of barrier	h_1, mm	h_2, mm	h_3, mm	P_k, GPa
Trinitrotoluene (TNT)	SiC	Dry	10.0	—	12.5	15.3
		Water saturated	11.0	—	16.0	19.2
		Monolith	—	—	12.0	25.7
	SiO_2	Dry	10.0	11.0	12,5	16.3
		Water saturated	16.0	18.0	21.0	20.4
TG-40	SiC	Dry	25.0	—	30.0	21.5
	SiO_2	Dry	22.5	—	23.7	22.0
		Water saturated	35.7	—	36.1	28.2

Source: From Merzhievskii et al. [9]. Reprinted with permission of Pleiades Publishing, Ltd.

Table 6.2 Results of experiments of the second series.

Powder # (grain size)	ρ, g/cm^3	h_1, mm	h_2, mm	h_3, mm	P_k, GPa
1 (1000–1250 μm)	1.24	27.0	—	28.0	16.1
2 (160–200 μm)	1.52	23.7	—	25.0	17.7
3 (3–5 μm)	0.97	25.0	27.0	28.0	13.9
4 (3–1250 μm)	1.92	22.0	—	22.5	22.0

Source: From Merzhievskii et al. [9]. Reprinted with permission of Pleiades Publishing, Ltd.

in [18–20]. The comparison showed an acceptable level of agreement between the calculated adiabats and the experimental curves.

In the second series of experiments [9], SiC powders of different fractional composition were used. Powder #1 was a grinding powder from black silicon carbide of grade 53c No. 100 F20 (grain size 1000–1250 μm); #2 was a grinding powder from green silicon carbide of grade 63c No.16 (M) (grain size 160–200 μm); #3 was a grinding powder from black silicon carbide grade 54c No. F1200 (M5) (grain size 3–5 μm); and #4 was a mixture of these powders in equal weight fractions (grain size 3–1250 μm). The mixture of powders of all fractions had the highest overall density, and the least dense was the powder with the smallest particle size. The experimental setup completely replicated the one used in the first experiments. The detonation transmission for TG-40/TG-40 pairs of charges was studied. The results of experiments of the second series are given in Table 6.2.

It should be noticed that it is difficult to compare these results with those of other authors. A proper comparison requires the mass, size, and type of explosives for active and passive charges to be identical. However, not just different countries but different laboratories in the same country may be using completely different values for these parameters. This leads to the fact that only a qualitative comparison of the main characteristics obtained in the experiments is possible. For practical purposes, it is useful to compare the weights of obstacles having a critical thickness for the transmission of detonation. Such a comparison of the obtained data with each other and with the results from [7] is shown in Table 6.3.

Considering the damping properties of the media studied, it should be noted that porous media are effective absorbers of energy under the influence of dynamic loads. As shown earlier [1], high-modulus dispersed media are capable of transforming a shock wave into an isentropic compression wave and desensitizing the explosive [21–23].

It can be argued that there exists a minimum pressure value, below which the explosive is not initiated. In our experiments, the initial pressure at the front of the shock wave entering the inert barrier is subsequently decreasing approximately according to an exponential law [24]. Unfortunately, the available data is insufficient to determine the

Table 6.3 Comparison of weight parameters for obstacles of critical thickness.

Material	Density, g/cm^3	Fractional composition, μm	Critical thickness, mm	Weight indicator, g/cm^3·mm
SiO_2	1.85	1000–100	23.1 ± 0.6	42.7
SiC	1.87	100–50	27.5 ± 2.5	51.4
	1.24	1250–1000	27.5 ± 0.5	34.1
	1.52	200–160	24.0 ± 0.5	36.5
	0.94–0.99	5–3	26.7 ± 1.0	26.4
	1.92	1250–3	22.3 ± 0.5	42.8
Steel	7.84	—	≈26.8	≈210
Aluminum	2.70	—	≈40	≈108
Polymethyl methacrylate (PMMA)	1.20	—	≈40	≈48
Water	1.00	—	≈40	≈40

Source: From Merzhievskii et al. [9]. Reprinted with permission of Pleiades Publishing, Ltd.

exponential factor, which depends significantly on both the porosity of the medium and the particle size of the powder.

Thus, high-modulus dispersed media have an advantage in the weight of the protective wall compared to metal (steel, aluminum) ones. For the same fractional compositions, ordinary sand possesses the same protection level as silicon carbide. Powders of silicon carbide with smaller particles have worse protective properties than sand. The best results according to the weight factor were obtained for fine silicon carbide powder. However, considering its high cost and complexity in handling, it cannot be considered as a promising material for practical applications.

6.2 Applications of the Protective Properties of Ceramic Materials

Excellent protective properties of ceramic materials allow using them as armor. Such armor, not yielding in effectiveness to armored steels, has a significantly lower weight. The development of ceramic armor, therefore, attracts a lot of attention, which is evidenced, for example, by the publication of 11 issues of international conference proceedings on ceramic and composite materials devoted to this topic [25].

A study of the penetration of impactors into ceramic barriers has shown that a high level of penetration resistance is achieved when impact velocities exceed the limiting rate of growth of brittle cracks [26].

Only a small number of publications have been devoted to the study of the high-speed impact on porous media [27]. The behavior of porous ceramics at high-speed impact was studied in [28]. In general, these results allow us to evaluate the ballistic resistance of porous materials.

These protective properties can be used for the development of containers and devices for the transport and storage of hazardous substances, including explosive materials. For instance, based on the research carried out by the authors of [3], a design of a container is proposed, in which the free space is filled with aluminosilicate microspheres with a diameter of 20–315 μm (POLIGRAN-2). It was also mentioned that glass powders and sand can be used in place of these microspheres, which is in line with the results of our studies. In [1], the use of ZrO_2, MgO (chamotte), and Al_2O_3 (corundum) powders as mechanical and thermal damping was justified.

The results of our research were used as the basis for the development of a container to increase the safety during transportation and storage of explosive materials, which uses ceramic interlayers made of materials, where the speed of sound exceeds the detonation velocity in the explosive [8].

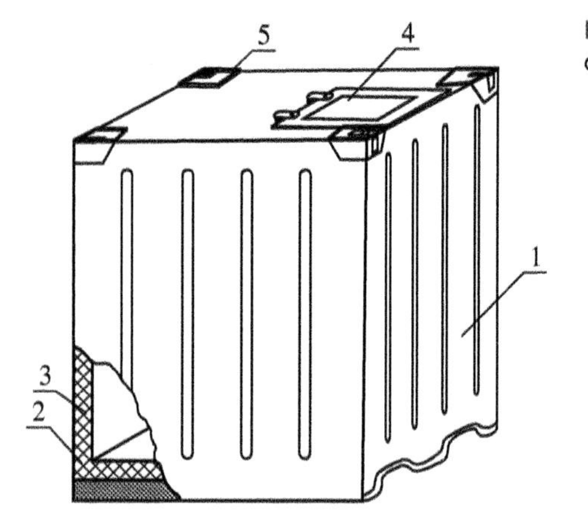

Figure 6.4 General view of the container.

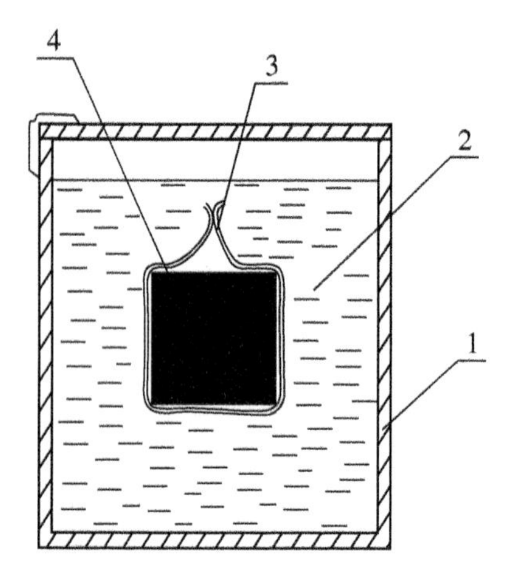

Figure 6.5 Another variant of the container.

The general view of the container is shown in Figure 6.4, where 1 is the outer shell, 2 is the intermediate protective layer, 3 is the internal case, 4 is the hatch for loading and unloading, and 5 is the lifting devices.

Another type of a container for storage and transportation of explosive materials was proposed in [12] (Figure 6.5). This container comprises an outer casing (1), an intermediate protective layer (2), and an inner casing (3), in which the explosive material (4) is placed. The intermediate protective layer is made of a high-modulus dispersion medium consisting of ceramic or glass particles (for example, silicon carbide, alumina, boron nitride, glass, sand, etc.), and the spaces between them are filled with a nonflammable inert medium (for example, water or air). Foamed or highly porous ceramics with similar properties can be used as well.

6.3 Summary

The results of the conducted studies can be used as a scientific justification for technical solutions for designing and manufacturing protective devices against emergency situations during the storage and transportation of explosive substances.

References

1 Ogorodnikov, V.A., Khokhlov, N.P., Yerunov, S.V. et al. (2003). Quasistatic and shock-wave behavior of perspective dampers of mechanical and thermal loads. *Proceedings of the international conference 'V Khariton Thematic Scientific Readings'*. Sarov, Russia (17–23 March 2003). Sarov: RFNC. [in Russian].

2 Meshkov, E.E. (2006). On the mechanisms of explosive load reducing in a closed volume with the help of layers of granular medium. *Proceedings of the International Conference 'VII Khariton Thematic Scientific Readings'*. Sarov, Russia (14–18 March 2005). Sarov: RFNC. [in Russian].

3 Drennov, A.O., Burtseva, O.A., and Gerasimenko, V.F. (2007). Improved container to prevent emergencies. *Proceedings of the International Conference 'IX Khariton Thematic Scientific Readings'* Sarov, Russia (12–16 March 2007). Sarov: RFNC. [in Russian].

4 Gelfand, B.E. and Silnikov, M.V. (2006). *Barothermic Action of Explosions*. St. Petersburg: Asterion [in Russian].

5 Ivanov, A.G., Fedorenko, A.G., and Syrunin, M.A. (1995). On the possibility of increasing the safety of nuclear weapons. *Combustion, Explosion and Shock Waves* 31 (2): 273–274.

6 Burlo, E. (1934). *Detonation Through Influence*. Moscow: Dzerzhinsky Artillery Academy [in Russian].

7 Orlenko, L.P. (ed.) (2002). *Physics of Explosion*. Moscow: Fizmatlit [in Russian].

8 Balagansky, I.A., Zorin, S.V., and Kaplouhov, V.M. (1995). Container for transportation and storage of explosive materials. Russian Federation Patent 2094751, filled 11 January 1995 and issued 27 October 1997.

9 Merzhievskii, L.A., Balaganskii, I.A., Matrosov, A.D. et al. (2012). Detonation transmission through high-modulus dispersed media. *Combustion, Explosion and Shock Waves* 48: 709–712.

10 Balagansky, I.A., Matrosov, A.D., and Stadnichenko, I.A. (2008). High-modulus heterogeneous dispersive media as a material of protective shells. *Proceedings of 3 Russian Conference 'Problems of Explosion Protection and Counter-Terrorism'*. Saint-Petersburg, Russia

(20–21 May 2008). Saint-Petersburg: St. Petersburg State University of Emergency Situations. [in Russian].

11 Balagansky, I.A., Matrosov, A.D., and Stadnichenko, I.A. (2008). High-modulus heterogeneous dispersive media as a material of protective shells. *Problems of Defensive Technologies. Series 16: 'Technical Means of Countering Terrorism'* 7–8: 64–69. [in Russian].

12 Balagansky, I.A., Matrosov, A.D., Stadnichenko, I.A. et al. (2008). Container for transportation and storage of explosive materials. *Proceedings of the 4th International Conference 'Technical Means of Counteracting Terroristic and Criminal Explosions'*. St. Petersburg, Russia (21–23 October 2008). Saint-Petersburg: St. Petersburg State University of Emergency Situations. [in Russian].

13 Balagansky, I.A., Matrosov, A.D., and Stadnichenko, I.A. (2008). Transmission of detonation through high modulus heterogeneous media. *Proceedings of the 6th All-Russian Scientific Conference Devoted to the 130th Anniversary of TSU and the 40th Anniversary of the Research Institute of Applied Mathematics and Mechanics 'Fundamental and Applied Problems of Modern Mechanics'*. Tomsk, Russia (30 September–2 October 2008). Tomsk: Tomsk State University [in Russian].

14 Dianov, M.D., Zlatin, N.A., Mochalov, S.M. et al. (1976). Shock compressibility of dry and water-saturated sand. *Pis'ma v Zhurnal tekhnicheskoĭ fiziki* 2 (12): 529–532. [in Russian].

15 Lagunov, V.A. and Stepanov, V.A. (1963). Measurement of dynamic compressibility of sand at high pressures. *Journal of Applied Mechanics and Technical Physics* 1: 88–96. [in Russian].

16 Afanasenkov, A.N., Bogomolov, V.M., and Voskoboinikov, I.M. (1969). Generalized shock Hugoniot of condensed substances. *Journal of Applied Mechanics and Technical Physics* 10: 660–664.

17 Alekseev, Y.L., Ratnikov, V.P., and Rybakov, A.P. (1971). Shock adiabats of porous metals. *Journal of Applied Mechanics and Technical Physics* 12: 257–262.

18 Trunin, R.F. (ed.) (2001). *Experimental Data on Shock Compression and Adiabatic Expansion of Condensed Matter Experimental Data on Shock Compression and Adiabatic Expansion of Condensed Matter*. Sarov: RFNC-VNIIEF.

19 Marsh, S.P. (ed.) (1980). *LASL Shock Hugoniot Data*. Berkeley: University of California Press.

20 McQueen, R.G., Marsh, S.P., Taylor, J.W. et al. (1970). Chapter VII – the equation of state of solids from shock wave studies. In: *High-Velocity Impact Phenomena* (ed. R. Kinslow), 294–415. New York: Academic Press.

21 Balagansky, I.A. and Gryaznov, E.F. (1994). Desensitization of RDX-charges after preshocking by compression wave in SiC-ceramic rod.

Proceedings of International Conference on Combustion 'Zel'dovich Memorial', Moscow, Russia (12–17 September 1994). Moscow: Russian Section of the Combustion Institute.

22 Balagansky, I.A., Balagansky, A.I., Razorenov, S.V. et al. (2005). Evolution of shock waves in silicon carbide rods. *Proceedings of the International Conference 'VII Khariton Topical Scientific Readings'*, Sarov, Russia (14–18 March 2005). Sarov: RFNC. [in Russian].

23 Balagansky, I.A., Balagansky, A.I., Razorenov, S.V. et al. (2006). Evolution of shock waves in silicon carbide rods. *Proceedings of the 14th APS Topical Conference on Shock Compression of Condensed Matter*, Baltimore, USA (31 July–5 August 2005). Melville: AIP Publishing.

24 Voskoboinikov, I.M. and Dolgborodov, A. Yu. (1986). Attenuation of shock waves in heterogeneous media. *Proceedings of the VIII All-Union Symposium on Combustion and Explosion*, Tashkent, USSR (13–17 October 1986). Chernogolovka: IPCP RAS. [in Russian].

25 LaSalvia, J.C. (ed.) (2015). *Advances in Ceramic Armor XI: A Collection of Papers Presented at the 39th International Conference on Advanced Ceramics and Composites*. Hoboken, NJ: Wiley.

26 Kozhushko, A.A., Rykova, I.I., and Sinani, A.B. (1992). Resistance of ceramics to projectile penetration at high interaction velocities. *Combustion, Explosion and Shock Waves* 28: 84–88.

27 Merzhievskii, L.A. and Chistyakov, V.P. (2014). High-velocity impact of steel particles on targets made of porous copper. *Combustion, Explosion and Shock Waves* 50: 498–500.

28 Miller, J.E., Bohl, W.E., Christiansen, E.L. et al. (2013). Ballistic performance of porous-ceramic, thermal protection systems. *Procedia Engineering* 58: 584–593.

7

Structure of the Materials Loaded Using Explosion Systems with High-Modulus Components

7.1 Materials Behavior at High Strain Rate Loading

The behavior of the material under high-speed loading may be different from its behavior under "classical," i.e. relatively slow methods of loading. The threshold value of the strain rate, which determines the boundary between static (quasistatic) and dynamic methods of influence on materials, is chosen rather arbitrarily. At the same time, most researchers believe that the processes of plastic deformation should be attributed to the dynamic in those cases when the inertial effects begin to have a significant impact on the behavior of the material.

In a review [1] Field et al. noted that the strain rate values achieved in real conditions are in the range of 10^{-8} s^{-1} to 10^{8} s^{-1} (Figure 7.1). Thus, they differ by 16 orders of magnitude. The lower range of strain rates is typical, for example, for creep processes, while the upper range is for deformation that occurs during the propagation of shock waves. The strain rate of 1 s^{-1} was chosen as a arbitrary boundary between the quasistatic and dynamic modes of deformation.

In most cases, the extremely high strain rates corresponding to the right edge of the diagram are associated with explosive technologies. Buijs noted that the use of an explosion in the materials fabrication is usually associated with the processes of explosion welding or cladding [2]. However, a number of other technologies based on explosive loading are used in industrial production, for example, explosion forming, explosion hardening, explosive compaction of powders, detonation spraying of coatings, etc.

Some of the most well-known technologies associated with the use of an explosion are considered in the well-known works of Meyers *Dynamic*

Explosion Systems with Inert High-Modulus Components: Increasing the Efficiency of Blast Technologies and Their Applications, First Edition. Igor A. Balagansky, Anatoliy A. Bataev, and Ivan A. Bataev.

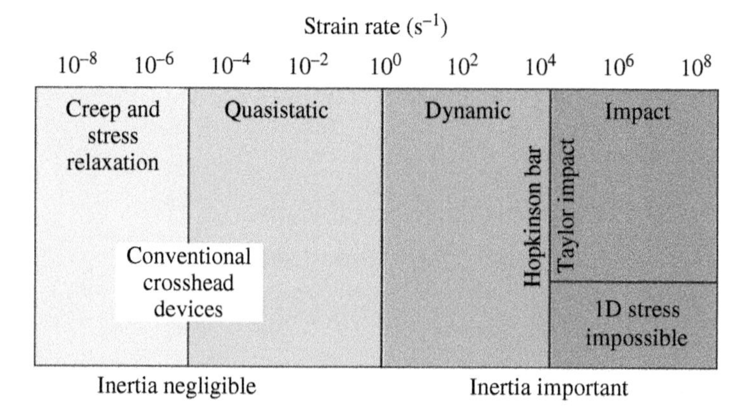

Figure 7.1 The range of strain rates existing in various industrial processes divided into several characteristic areas. (*Source:* From Field et al. [1]. Reprinted with permission of Elsevier.)

Behavior of Materials [3], Zukas and Walters *Explosive Effects and Applications* [4], and Selivanov et al. [5]. A brief analysis of mentioned explosive technologies in various sources is presented in Table 7.1.

Although the technologies listed in the table are being actively implemented in various industries, the structure of the materials obtained with their use has not yet been studied in sufficient detail. This is due to several reasons. One of them is associated with a small number of laboratories performing work with explosives. Moreover, in recent years, environmental requirements, as well as requirements for the safety of storage, transportation, and use of explosive materials, have already been significantly tightened. This means that the possibilities of conducting mass experiments for scientific purposes are currently significantly limited.

Besides, structural transformations in materials exposed to explosive loading can only be reliably observed using modern analytical equipment, in particular, high-resolution transmission electron microscopes (TEM), scanning electron microscopes (SEM) equipped with highly sensitive spectrometers, etc. allowing accurate investigation of the composition of local areas of the material. A large number of experimental data obtained during fist decades after emerging of explosive technologies (~ mainly in 1950th and 1960th) were received using rather simple analytical equipment, the capabilities of which were significantly inferior to modern instruments.

But the largest difficulty is related to the short duration of explosive loading. The structural transformations caused by the explosion occur for extremely short periods of time, usually in the range from a few

Table 7.1 Analysis of several monographs on explosive technologies.

	Meyers [3]	Zukas and Walters [4]	Selivanov et al. [5]	Cooper [6]
Explosive welding	+	+	+	+
Application of shaped charges for military purposes	+	+	+	+
Application of shaped charges for peaceful purposes	+	+	+	+
Breaking through barriers (armor)	+	+	+	+
Explosive forming	+	−	+	−
Explosive hardening	+	−	+	−
Explosive compaction of powders	+	−	+	−
Shock-induced chemical synthesis	+	−	+	−
Shock-induced phase transformations	+	−	+	−
Detonation spraying	−	−	+	−
Peaceful nuclear explosions	−	−	+	−

nanoseconds to several microseconds. The *in situ* visualization and analysis of these processes is very difficult from a technical point of view. For this reason, many fundamental ideas on the evolution of materials structure in the processes of explosive loading were obtained mainly based on characterization of samples after the completion of loading (i.e. based on *postmortem* study of samples) or based on the results of numerical simulation.

The behavior of a material during the plastic deformation, in particular its plastic flow, plastic hardening, fracture, etc., is largely determined by its structure. In accordance with the currently dominant theoretical concepts, the strength and plasticity of metallic materials are associated with their dislocation structure. Many researchers (see, for instance, studies of Tushinskij [7], Mecking and Kocks [8], etc.) developed the idea that the yield strength of materials is determined by a combination of the following hardening mechanisms:

- The resistance that the crystal lattice renders to the movement of dislocations (i.e. slowing down of dislocations by the Peierls barriers).
- Barriers built of randomly located dislocations and their clusters.

- Hardening by atoms of alloying elements dissolved in the matrix.
- Hardening due to precipitation of fine particles of other phases.
- Hardening due to the high-angle boundaries (i.e. grain boundaries).

The strength of materials can also be represented as a combination of athermal (σ_a), frictional (σ_0), and thermally activated (σ_ρ) members [9]. The last two terms essentially depend on the temperature and strain rate:

$$\sigma = \sigma_a + \sigma_0\,(\dot{\varepsilon}, T) + \sigma_\rho\,(\dot{\varepsilon}, T, \varepsilon). \tag{7.1}$$

Behavior of materials may differ significantly at different loading rates. It is frequently believed that increasing the strain rate has the same effect on the yield stress as lowering the temperature. Gray noted that the decrease in flow stress due to thermal activation is one of the most important reasons explaining the change in the deformation behavior of materials due to variation of strain rates [9].

It should be emphasized that the effect of strain rate on the evolution of the microstructure and characteristics of strength and ductility is a unique feature of each particular material. To date, an extensive amount of structural studies of body-centered cubic (bcc), face-centered cubic (fcc), and hexagonal close packed (hcp) metals subjected to plastic deformation at high speeds has been performed. At the same time, the literature contains a significantly smaller number of similar studies of materials with a more complex structure. In particular intermetallics or ceramic-based alloys subjected to explosive loading are poorly studied.

There are several microstructural peculiarities that are typically observed in materials with an increase in the strain rate: more uniform distribution of dislocations, suppression of the formation of dislocation cells, a decrease in cell size, an increase in the misorientation angle between neighboring cells, and an increase in the number of dislocations observed inside the cells [10, 11]. One of the reasons for the observed microstructures is associated with a decrease in the time available for dislocations rearrangement.

Shock wave loading may be considered a limiting case of dynamic loading of materials (Figure 1.1). In contrast to the "slow" plastic waves observed at relatively soft dynamic loading, a sharp jump in pressure, density, temperature, and other parameters occurs at the shock wave front.

In [12], Meyers and Murr noted that there are no fundamental differences between the dislocation structure of shock-loaded and "slowly" loaded materials. However, some differences do exist. The dislocations cells are frequently observed in the shock-loaded samples. In contrast to "slowly" loaded samples, the density of dislocations inside the cells in the shock-loaded samples is much higher.

Among all the parameters, which characterize the shock loading, the shock pressure has the greatest influence on the dislocation structure. Up to approximately 100 GPa the residual dislocation density varies in proportion to the square root of pressure. At higher pressures, this law is violated, which is probably related to the heating associated with the shock loading. The duration of the shock pulse has much smaller effect on the dislocation structure of the material. An increase in the pulse duration leads to increase of the time available for rearrangement of dislocation structure. Thus, an increase in duration of the shock pulse leads to the formation of dislocation cells with more distinguishable walls.

One of the first studies devoted to characterization of the dislocation structure of materials subjected to the shock loading was the one published by Smith [13]. It was assumed that the boundary between the shock wave and the uncompressed material should consist of an array of dislocations, which are necessary to provide a geometric mismatch between compressed and uncompressed grids. In accordance with this assumption, the dislocation density at the shock front should be by three to four orders of magnitude greater than the observed residual dislocation density of experimental samples subjected to shock loading. Based on this observation, Smith supposed that behind the shock wave, the dislocation "sinks," which absorb excessive dislocations, should also move.

The model proposed by Smith was modified in studies of Hornbogen [14]. In accordance with his hypothesis, the dislocation loops are formed during the motion of the shock wave. The edge components of loops have to move with the shock front velocity, providing the necessary geometric mismatch, while the screw components remain in the shock wave and are stretching behind the edge component. Both of these models have been criticized in study [12] of Meyers and Murr, who showed that the motion of a large number of dislocations with such high velocities would lead to the heating of the material to extremely high temperatures, which is not observed in real experiments. Due to this reason, Meyers proposed a model of nucleation and motion of dislocations in shock waves [15, 16] (Figure 7.2), based on the following assumptions:

1) Dislocations nucleate homogeneously at the front of the shock wave or immediately behind it as a result of the deviator components of the stress tensor (Figure 7.2b).
2) Dislocations move for short distances with subsonic speeds.
3) As the shock front moves forward, new surfaces of the nucleation of dislocations are created.

In a review study [17], Armstrong and Walley noted that under conditions of shock loading, the strength and ductility of materials are controlled not by the dislocations mobility, but by the rate of nucleation of

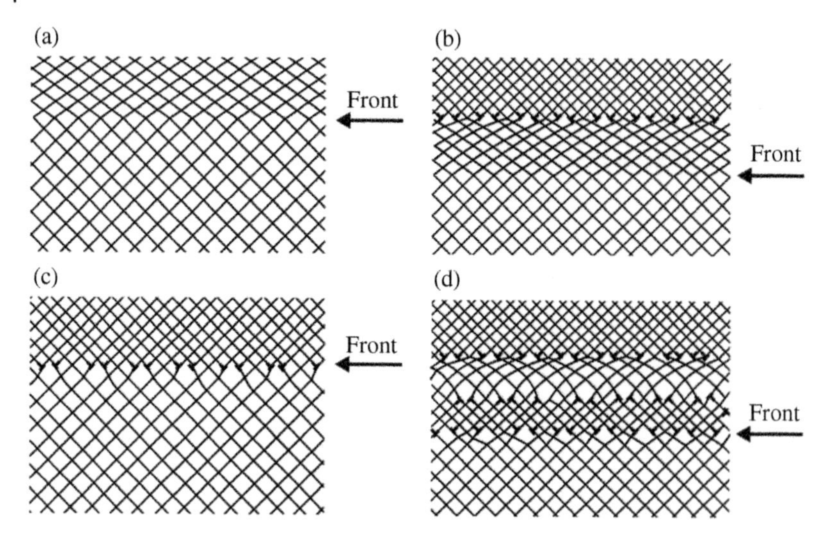

Figure 7.2 Propagation of the shock front and nucleation of dislocations in it according to model of Meyers. (a) Lattice mismatch at the shock front. (b) Formation of the dislocations. (c) Propagation of the shock front. (d) Formation of new set of dislocations. (*Source:* From Meyers [15]. Reprinted with permission of Elsevier.)

defects on the shock front. According to Weertman, the movement of dislocations with velocities close to the speed of sound in a crystal requires the shear stresses approaching the theoretical shear strength of the material [18]. At the same time, in cases where the shear stresses in the shock waves exceed the theoretical crystal strength, the dislocations can move with arbitrarily high velocities. At the same time in study [12], Meyers and Murr claimed that dislocations cannot move with a speed equal to or greater than the speed of sound in crystals.

Another feature of materials behavior under high-strain-rate plastic deformation is an increased tendency to deformation twinning. In particular, in bcc metals, the stresses required for a slip of dislocations increase sharply with decreasing temperature, while the critical stresses for twinning decrease slightly or remain constant with decreasing temperature [9, 12, 19] (Figure 7.3). Thus, with a decrease in temperature, as well as with an increase in the strain rate, in many bcc metals, the mechanism of deformation twinning is activated, and the dislocation slip is suppressed.

Some metals may experience phase transformations under the action of shock loading. In accordance with thermodynamic concepts, increase in pressure leads to stabilization of phases with higher density. For example,

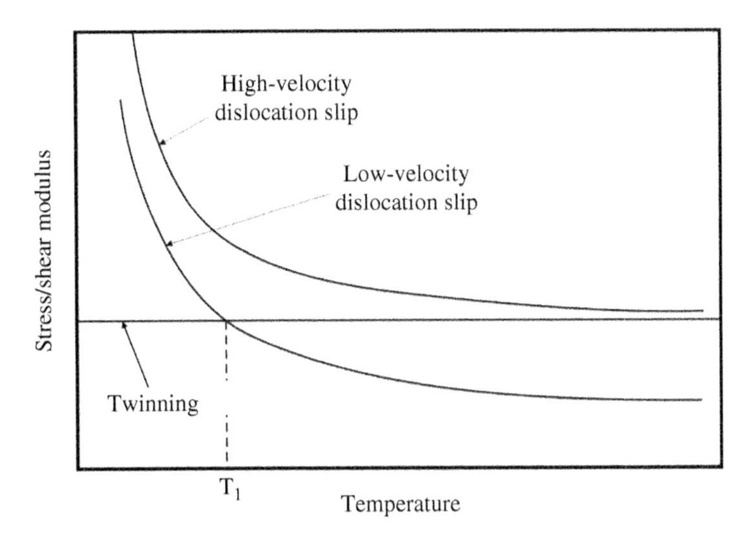

Figure 7.3 The effect of temperature and strain rate on stresses of deformation twinning and dislocation gliding. (*Source:* From Meyers and Murr [12]. Reprinted with permission of Springer Nature.)

it is well known that at pressure of more than 13 GPa, α-Fe having bcc structure transforms to a denser ε-iron with hcp structure [20, 21]. After the passage of shock waves, the reverse $\varepsilon \rightarrow \alpha$ transformation should develop in iron leading to a formation of a significant number of boundaries, resembling boundaries of twins. In many materials, after the passage of shock waves, the reverse transformation does not completely occur, and shock loading is accompanied by the formation of a complicated structure containing residual metastable phases. Shock-induced $\alpha \rightarrow \omega$ transformations in titanium and zirconium are good examples of such phenomenon.

One of the manifestations of shock wave loading is a significant increase in the concentration of point defects [3]. For example, in [22] it was noted that the concentration of vacancies and interstitial impurities after shock loading is four times higher than after cold rolling. The vacancies created during the passage of shock waves through the metals are reliably identified by TEM methods, since they lead to the formation of specific dislocation loops. The mechanisms of the appearance of vacancies in the case of shock wave loading do not have fundamental differences as compared with their formation under slow loading and it may be considered as the intersection of screw or mixed dislocations.

Explosive loading frequently leads to localization of plastic deformation. One of the most striking manifestations of the deformation localization under conditions of high-strain-rate deformation is the formation of adiabatic shear bands.

In the early stages of the deformation of many materials, the plastic flow occurs uniformly throughout the volume. However, under certain conditions, the behavior of the material can suddenly change, and plastic deformation concentrates in narrow layers of the sample [23]. On the polished cross sections of such materials, one may observe specific areas with significant plastic flow that are typically called "shear bands." The localization of plastic flow and the formation of shear bands can occur even under moderate strain rate, but this phenomenon is especially typical of high-strain-rate loading.

At high strain rates, the material in the shear bands can almost instantly heat up to high temperatures since the heat generated during the deformation process does not have time to be transferred into the surrounding layers. Due to this reason, the bands formed in the process of high-strain-rate loading are called "adiabatic shear bands." Nevertheless, it should be noted that in some studies, for example [24], a lot of attention is paid to the heat exchange between the shear bands and the surrounding material, and the relationship between the critical strain rates necessary for the appearance of bands and the thermal diffusivity of the metals was observed.

In accordance with the ideas of Zener and Hollomon [25], who observed the formation of adiabatic shear bands during metal punching, if some part of the material deforms more intensively than the neighboring material and softens as a result of this process, then the deformation localizes in it, while the surrounding material volumes are not involved in the deformation. Such mechanism explaining the formation of the adiabatic shear bands has not undergone fundamental changes up to the present time. Zener and Hollomon assumed that the material in the adiabatic shear bands formed during the deformation of steel underwent martensitic transformation upon subsequent cooling.

Hargreaves and Werner described the adiabatic shear band as a result of thermomechanical instability, appearing in cases where the effect of strain hardening is lower compared with softening due to the formation and growth of voids (geometric softening) and thermal softening during heat generation [26].

The increased attention of the scientific community to the adiabatic shear bands is related to the fact that they lead to simplified material fracture under high-strain-rate deformation, which is a problem for developers of armor materials. Besides, the adiabatic shear bands retained inside the sample lead to its embrittlement.

The width of the adiabatic shear band is typically in the range from few micrometers to several tens of micrometers, with the values of strain in their core reaching 100 and the values of strain rate exceeding 10^6 s^{-1} [23]. It is also important to note that the cooling rates of the material in the adiabatic shear bands can reach values higher than 10^6 K/s. Microhardness measurements in the directions crossing the adiabatic shear bands revealed that depending on alloy composition the material in the most deformed areas can be harder or softer than the surrounding microvolumes [27–30].

The width of the adiabatic shear band (δ) depends on the shear rate $\dot{\gamma}$, the temperature reached in the band (θ_b), the thermal conductivity of the material (λ), and the shear stresses τ_b in accordance with the approximate equation [31]

$$\delta = \left(\frac{\lambda \theta_b}{\tau_b \dot{\gamma}} \right)^2. \tag{7.2}$$

Thus, it can be assumed that the narrower is the observed shear band, the lower was the temperature, and the greater were the stress and strain rate in the process of its growth. All other things being equal, narrower shear bands are formed in materials with a lower thermal conductivity. A more rigorous analysis of the influence of the materials properties and deformation modes on the width of the adiabatic shear bands is presented in studies of Grady and Kipp [32, 33].

The problem associated with the experimental measurement of temperature in adiabatic shear bands at the stage of their formation is extremely difficult from experimental point of view. According to some experimentally obtained estimates, the temperature reached in adiabatic shear bands in steel at the stage of their propagation can be as high as 1100 °C [34]. According to different studies the temperature values can be either lower (~900 °C [35]) or much higher (up to the melting temperature for some alloys [36]).

The technologies based on high-strain-rate deformation of materials, which are currently used in real industrial production, as a rule, lead to complex stress and strain states. It should, however, be noted that most of the basic research devoted to the analysis of the features of structure formation under high-speed loading is carried out under simplified conditions, which try to achieve a plain stress or plain strain states of materials. For this purpose, materials are subjected to tests on a Split–Hopkinson pressure bar, falling weight impact test, Taylor impact test, or shock loading by flat collision of plates [37].

7.2 Postmortem Investigation of Materials Structure for Indirect Evaluation of Explosive Loading

One of the most important issues related to the simulations of fast processes, such as processes of explosive loading, is to assess the validity of the calculated values. To measure necessary values in real experiments, one may use various sensors to record parameters of the loading process or visualize an explosion using high-speed cameras. A useful addition to these methods is application of metallographic analysis of "witness" samples.

It is known that certain values of pressure and temperature lead to specific microstructures that may frequently be unique for certain materials. The greatest amount of information in the analysis of microstructures can be obtained in cases where the material undergoes a significant amount of phase transformations, caused by both temperature changes and pressure changes. Titanium- and iron-based alloys can be distinguished among other materials, which are widely used in industry. The phase diagrams of pure titanium [38] and pure iron [39] in P–T coordinates are shown in Figure 7.4. From our point of view, iron-based alloys (primarily low-carbon steels) are particularly well suited to be used as witness samples due to a significant variety of pressure- and temperature-induced phase transformations and relatively low price. Another advantage of iron-based alloys is a significant amount of research devoted to study of microstructures arising in these materials during various combinations of thermal and deformation impacts. Thus, by comparison of the microstructure obtained in iron-based samples with a P–T phase

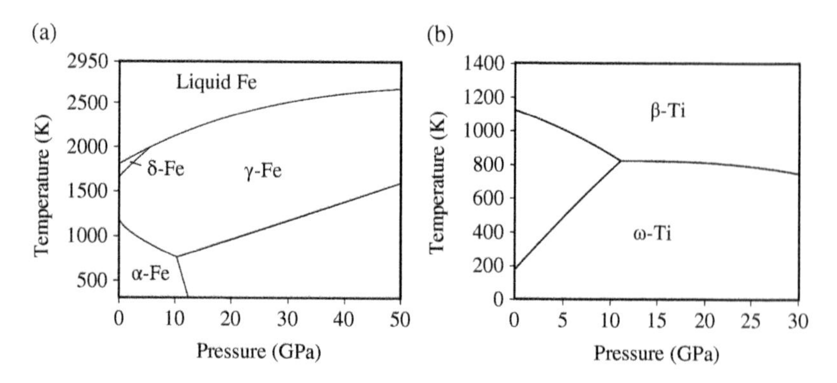

Figure 7.4 Pressure–temperature phase diagrams of iron and titanium for iron (a) (*Source:* From Dinsdale [39]. Reprinted with permission of Elsevier) and titanium (b). (*Source:* From Mei et al. [38]. Reprinted with permission of American Physical Society.)

diagram and published data on various microstructures arising in these alloys, one may frequently rather accurately estimate the loading parameters and compare them with the results of numerical simulation.

As an example of such an analysis, one may consider our experiment on high-velocity oblique collision of plates accelerated in a powder gun. In this experiment, the process of high-velocity impact welding of iron plates was studied. A detailed description of the experiment, numerical simulation, and analysis of microstructures are presented in Figure 7.5.

The scheme of the experiment is shown in Figure 7.5. Samples for welding had the shape of disks with a diameter of 38 mm and a thickness of 1 mm (Figure 7.5c). The flyer plate was attached to a cylinder made of polyethylene. The parent plate was placed at an angle of 15° with respect to the flyer plate (Figure 7.5d). The experiments were carried out in

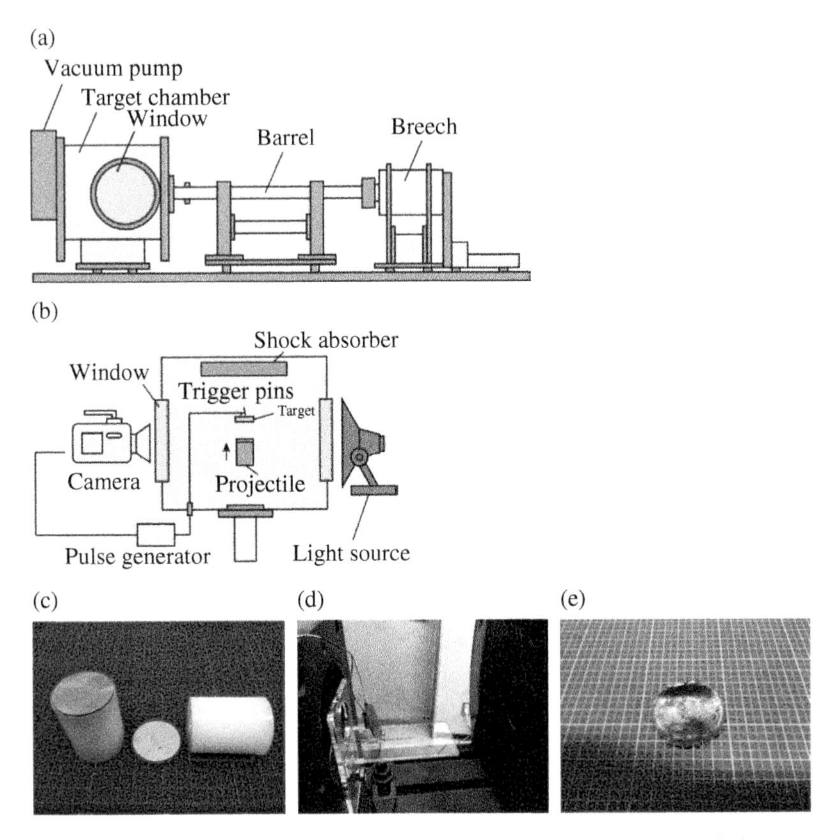

Figure 7.5 Scheme of experiment on high-velocity impact welding of steel plates (a and b), images of starting materials (c), initial arrangement of parent plate in powder gun chamber (d), and image of the welded sample (e).

vacuum. The collision velocity, as well as related phenomena, was monitored using a high-speed camera HPV-1 (Shimadzu Co., Ltd) at one million frames per second. Three impact velocities were implemented during experiments: 542, 735, and 1000 m/s.

A typical welded sample is shown in Figure 7.5e. It was cut along the direction of welding and then gridded and polished using sanding papers and diamond suspensions. The structure was studied by optical microscopy, SEM, and TEM.

ANSYS AUTODYN 17.0 software was used to analyze the flow of materials during high-velocity collisions. To analyze the conditions corresponding to the shock wave, the "shock" equation of state was used. The parameters of the equation of state that corresponded to the low carbon steel studied were obtained from the standard database of AUTODYN (Table 7.2). The dependence of the material strength on the loading conditions was analyzed using Johnson–Cook equation (Table 7.3).

Table 7.2 Parameters of "shock" equation of state for low-carbon steel (according to standard database of AUTODYN).

Parameter	Value
Initial density, g/sm^3	7.896
Gruneisen parameter, Γ_0	2.17
Parameter C_1, m/ms	4.596
Parameter S_1	1.49
Initial temperature, K	300
Specific heat c_0, kJ/gK	$4.52 \cdot 10^{-4}$
Thermal conductivity	0

Table 7.3 Parameters of Johnson–Cook equation for low-carbon steel (according to standard database of AUTODYN).

Parameter	Value
Shear modulus, GPa	81.8
Coefficient A, GPa	0.35
Coefficient B, GPa	0.275
Coefficient n	0.36
Coefficient C	0.002
Coefficient m	1
Melting temperature T_m, K	1811

To solve the equations of continuum mechanics written in differential form, the method of smoothed particle hydrodynamics (SPH) was used. This method is currently considered as an effective approach for solving problems associated with large plastic deformations.

Based on comparison of microstructures with data of simulation and the literature data, the pressure and temperature in the collision zone were evaluated.

Figure 7.6 shows several frames that were recorded during experiments on high-velocity impact of steel billets. Figure 7.6c–f shows a bright jet, which is formed soon after the beginning of the collision. This result is in a good agreement with most literature data on explosive welding, which assume formation of the so-called reentrant jet moving in front of the collision point.

From Figure 7.7 one may observe that in all cases of collision, wavy interfaces were formed (Figure 7.7a–c). The plastic strain gradually increased near the interface. In the vicinity of interface, the grains of ferrite are stretched along the waves (Figure 7.7d).

Careful examination shows presence of a dark-etching layer near the interface of the sample. This layer is clearly visible in the sample, obtained at collision velocity of 1000 m/s (Figure 7.7c), and it is much narrower in the sample, obtained at velocity 735 m/s (Figure 7.7b). In the sample, obtained at velocity of 542 m/s, this layer cannot be clearly distinguished using optical microcopy. Comparison of images, shown in Figures 7.7c, f

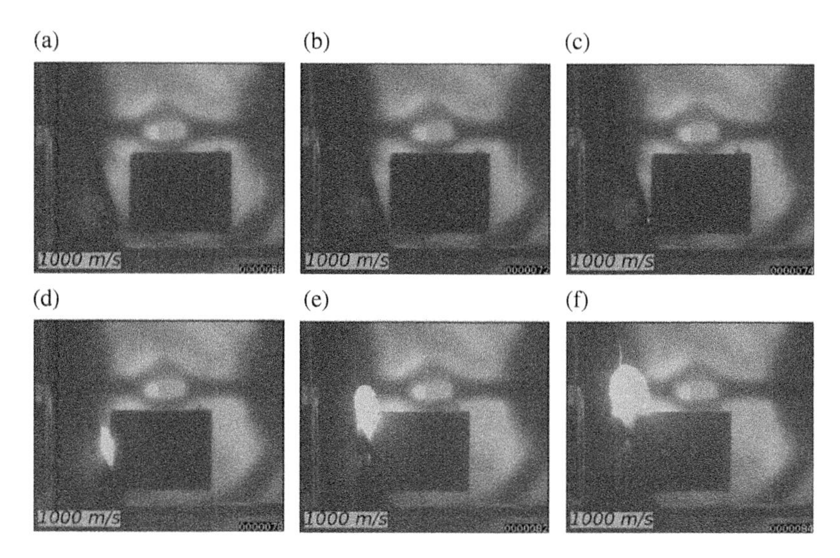

Figure 7.6 Collision of steel plates in powder gun at velocity 1000 m/s and at an angle of 15° recorded using high-speed video camera. (a–f) Correspond to different moment of time (as shown at right bottom corner).

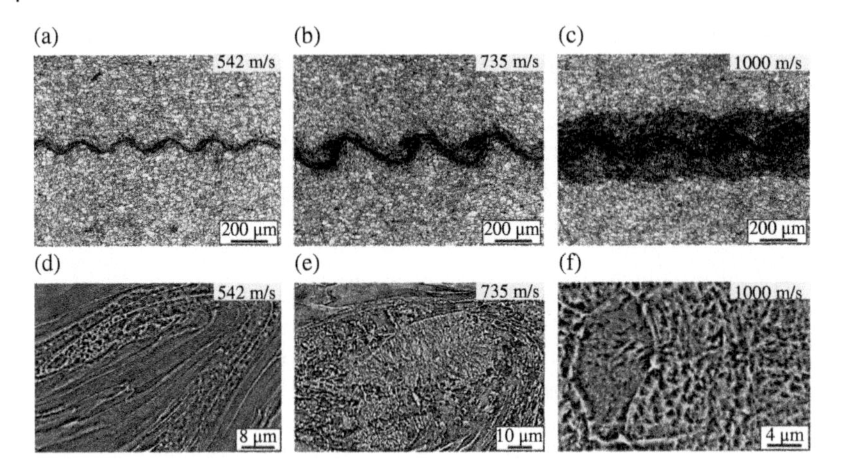

Figure 7.7 Images of interfaces obtained by high-velocity collision of plates in powder gun. (a and d) Flyer plate velocity 542 m/s, (b and e) flyer plate velocity 735 m/s, and (c and f) flyer plate velocity 1000 m/s.

Figure 7.8 TEM image of zone of $\alpha \rightarrow \varepsilon \rightarrow \alpha$ transformation in a sample welded at velocity 1000 m/s.

and 7.8 with results, obtained by Kheifets et al. [40] allows us to conclude that this layer is produced by $\alpha \rightarrow \varepsilon \rightarrow \alpha$ transformation.

An important feature of high-velocity impact welding is formation of a liquid phase at the interlayer boundary. These areas are frequently referred to as vortices due to specific twisting of metal fluxes. This phenomenon is particularly well seen when joining dissimilar alloys. In this study a specific twisting near the interface may be noticed due to the specific shape of ferrite grains. The zones of solidified liquid may be observed on lateral sides of the waves.

The TEM images of the sample obtained at velocity of 1000 m/s are shown in Figure 7.8. These images correspond to the dark zone of $\alpha \rightarrow \varepsilon \rightarrow \alpha$ transformation, which was previously discussed. Due to the specific diffusionless mechanism of $\alpha \rightarrow \varepsilon \rightarrow \alpha$ transformation, one may observe a specific lath structure.

Comparison of experimental results and simulated maps of strain and temperature distribution (Figure 7.9) shows remarkable coincidence of the interface shape. The location and shape of molten pockets was also clearly reproduced.

By comparison of pressure and temperature information received during numerical simulation with P–T phase diagram of iron, one can plot a map of phase state achieved by the metal due to the influence of high pressures and corresponding temperatures. This map is shown in Figure 7.10 for the sample, received at velocity of 1000 m/s. One may observe that the size and shape of the zone of ε-phase formation predicted based on simulation is just slightly bigger in comparison with dark-etching layer observed in experiments.

It can also be noticed that the time that is needed to decrease the pressure is much lower than time needed to cool down the interface to the initial state. For this reason, not only reverse $\varepsilon \rightarrow \alpha$ transformation but also $\varepsilon \rightarrow \gamma \rightarrow \alpha$ and other complex transformations may occur.

The critical pressure needed for $\alpha \rightarrow \varepsilon$ transformation is 11–13 GPa depending on the temperature. Based on the simulation it was observed that such high pressures are achieved only in a very narrow zone near the interface, which is in a good agreement with the fact that dark-etching layer is almost invisible in experiments with the velocities of 542 and 735 m/s.

Summing up the results presented in this section, we can note a good agreement between the calculated and experimental data, which, from our point of view, increases the degree of confidence in the calculated data. Thus, the comparison of the results of metallographic analysis and numerical simulation is an effective tool, the use of which helps to understand the nature of the phenomena that occur during the explosion.

7.3 Structure of Materials Loaded Under Conditions of Energy Focusing

This section presents the results of the study of steel witness samples used in experiments on energy focusing described in Section 2.3.

Steel disks with a diameter of 120 mm and a thickness of 12 mm made of hot-rolled low-carbon steel (0.18% C; 0.51% Mn; 0.16% Si; 0.03% P;

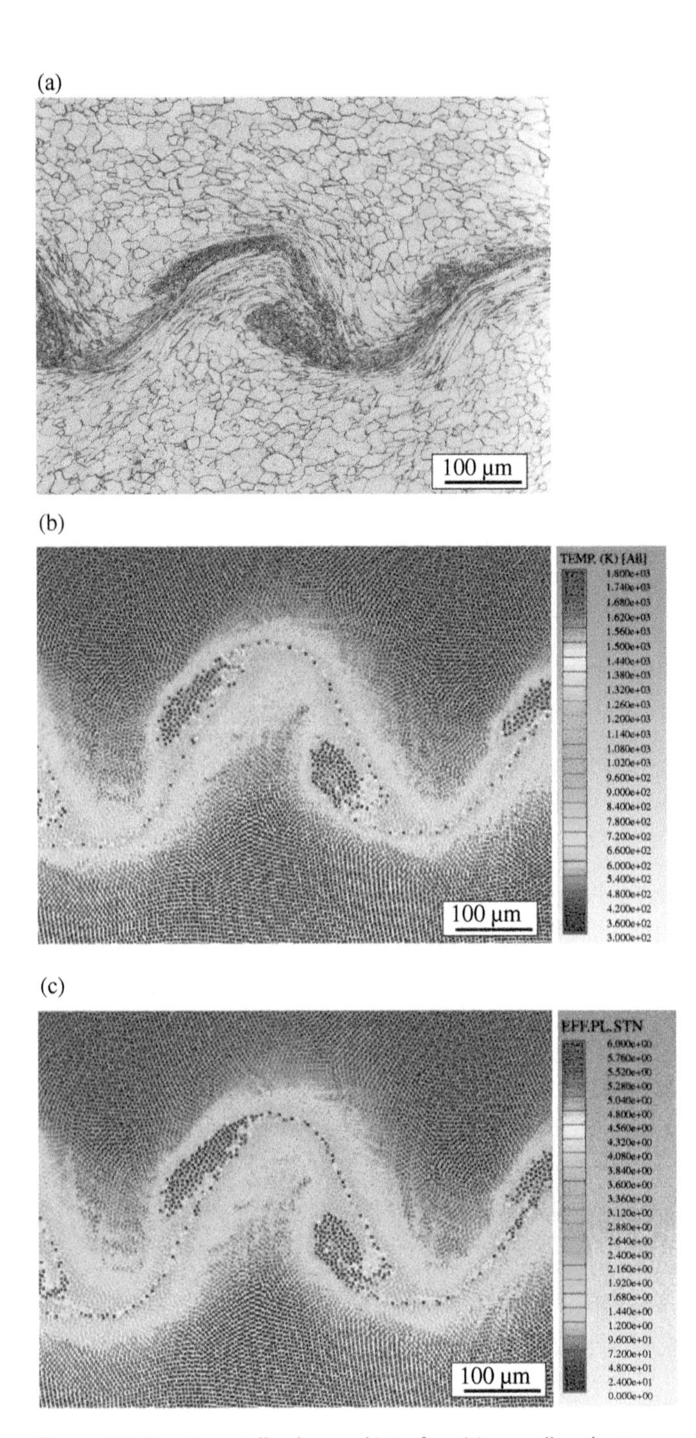

Figure 7.9 Experimentally observed interface (a), as well as the temperature distribution (b) and pressure (c), obtained by smoothed particle hydrodynamics (SPH) simulation.

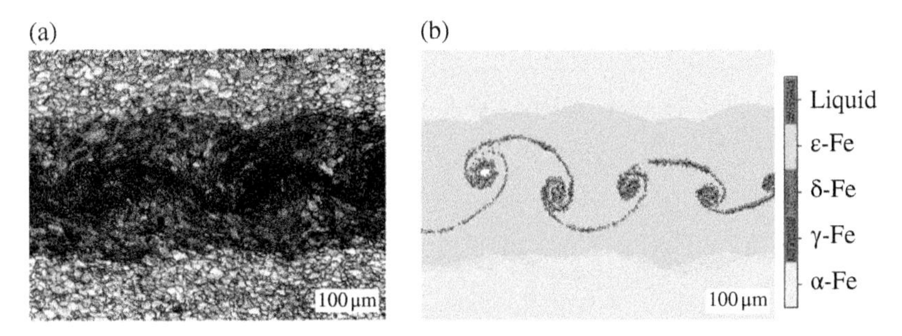

Figure 7.10 Comparison of experimental results (a) and predicted phase state of iron plates (b).

0.04% S) with a ferrite–pearlite structure were used as the witness samples. The average size of the ferritic grain was 28 μm. Lamellar pearlite was in the form of conglomerates of colonies, observed in the pictures in the form of dark (strongly etched) areas. The average size of the conglomerates was 42 μm. High-strain-rate loading was carried out using the phenomenon of energy focusing in an explosive system that includes high-modulus elastic elements as it was described in Section 2.3. Metallographic studies were performed using a Carl Zeiss AXIO Observer A1m microscope at magnifications in the range of ×25–1500. Characterization of tiny microstructural details was carried out using SEM (Carl Zeiss EVO 50 XVP microscope) and TEM (Tecnai G2 20 TWIN microscope). The microhardness of various zones was determined using a Wolpert Group 402 MVD instrument.

The disks loaded according to the scheme shown in Figure 2.16 acquired a cup-shaped form (Figures 7.11 and 2.25). Three zones that differ in the nature of the processes of plastic flow and material failure were observed in samples. The diameter of the central zone (zone I), in which the most pronounced structural transformations were observed, was approximately 40 mm. In the peripheral zone of the disk (zone III), no noticeable changes in the structure of ferrite and pearlite due to high-speed deformation were found. This zone had the shape of a ring with an outer diameter of 120 mm and an inner diameter of approximately 110 mm. The zone II, intermediate in location, had a gradient, explosion-deformed structure. The number of defects of deformation origin in zone II increased toward the center of the disk.

The main type of defects that form in the middle zone of the disk (zone II) with explosive loading was deformation twins. Most of them were formed in ferritic grains (Figure 7.12a and b). This mechanism of plastic flow is facilitated by external loading conditions: a high deformation rate and a relatively low temperature of the material during the

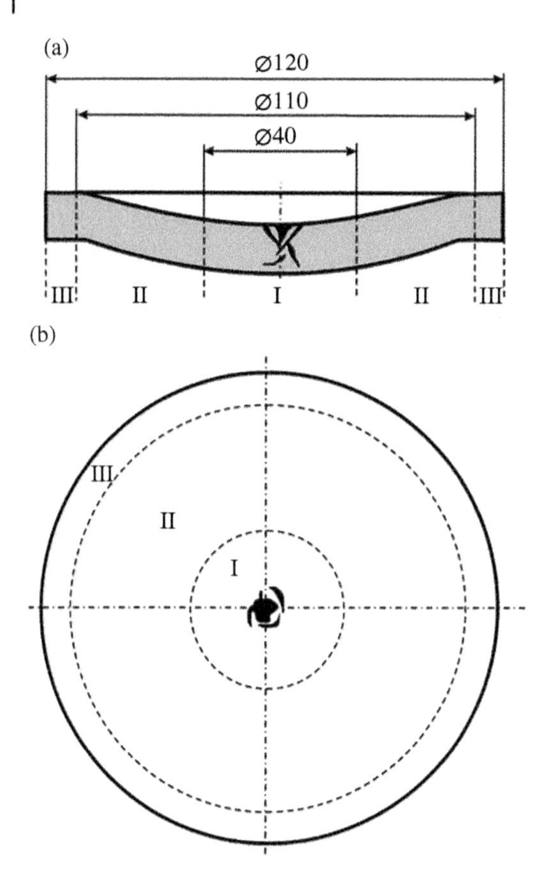

Figure 7.11 Scheme of the disk after loading using the scheme of energy focusing described in Section 2.3. (a) Front view and (b) top view.

deformation. When these conditions are met, mechanical twinning is manifested when the slip planes are unfavorably oriented with respect to the acting shear stresses [41]. For conditions of static and quasistatic loading of ferrite, the twinning phenomenon is not typical. In these cases, the deformation of the α phase develops according to the mechanism of dislocation slip.

The important feature of the twins is their clear crystallographic relationship with the initial matrix of the phase in which they were formed. In crystals with a bcc lattice, the processes of mechanical twinning develop in systems {112} <111> [41]. For this reason, the twins observed in ferrite look like sets of bands parallel to each other. Usually twins spread from one grain boundary to another. However, few twins were found that were interrupted inside the ferritic grain (Figure 7.12a).

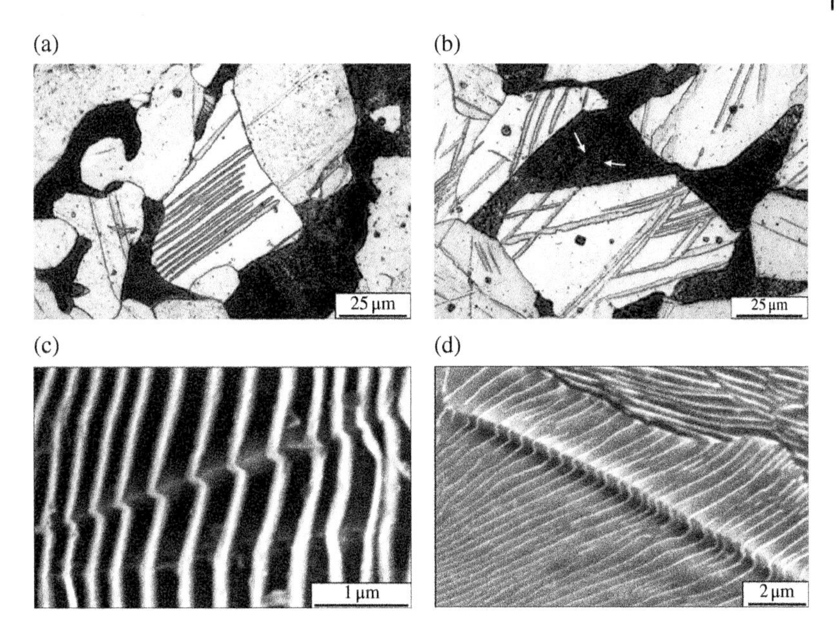

Figure 7.12 Images of moderately deformed zone II showing formation of multiple twins in grains of ferrite and pearlite colonies. (a and b) Light microscopy and (c and d) scanning electron microscopy.

The deformation twins in ferrite form relatively narrow layers in the original matrix. Their thickness, determined by TEM, is about 0.2–1 μm. Dozens of narrow twins parallel to each other were metallographically observed within the limits of one ferrite grain. The greatest number of twins (more than hundred) was observed in large ferritic grains (~60–70 μm in size). Thus, in the process of high strain rate deformation and formation of a system of twins parallel to each other, ferritic grains are divided into a number of areas with several variants of crystallographic orientation with respect to original α-phase (Figure 7.13). In some of the grains, one can observe two to three sets of twins parallel to each other. Under these conditions, the boundaries of twins of one orientation can serve as barriers to the propagation of twins of another orientation. Thus, twins of different orientations form a pattern in the form of characteristic, regularly intertwining "farms" (Figure 7.12b).

Using optical microscopy and SEM, we observed that twins can be formed not only in a single-phase material (ferrite) but also in the structure of lamellar perlite (Figure 7.12c and d). Twinning processes in heterophase materials are poorly studied. For this reason, the analysis twin's formation in perlite of low-carbon steel is of particular interest. In Figure 7.12b one can observe two parallel twins (marked by arrows),

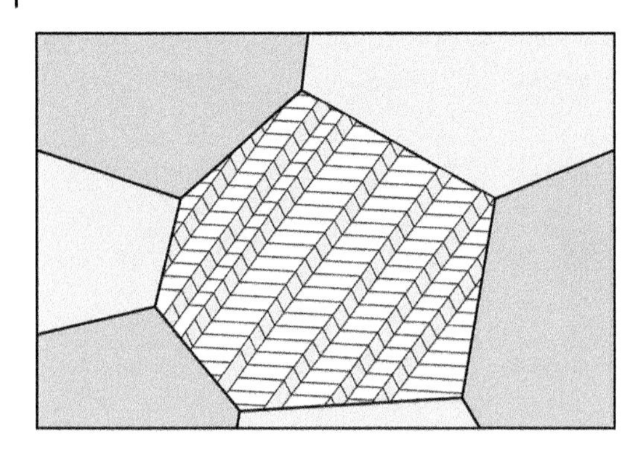

Figure 7.13 Scheme showing the change in the orientation of the crystallographic planes of ferrite in the formation of several deformation twins belonging to one system.

propagating through the grain of ferrite and having found their continuation in the neighboring pearlite colony. It can be assumed that the crystallographic orientation of the ferritic grain and the ferritic matrix of the neighboring pearlite colony coincides.

The most noticeable structural transformations in the studied steel were recorded within zone I. A crater with a diameter of 5–10 mm is formed directly in the center of the disk. The tempering colors observed on the surface of the material near this crater indicate the high temperature occurring during the deformation of a steel disk by an explosion. The formation of a crater is due to the phenomenon of energy focusing, which is associated with the initiation of detonation in a massive explosive charge pre-compressed by the leading wave and the convergence of the detonation wave to the axis of symmetry of the charge.

The number of defects occurring in zone I was much greater than in zone II. The main type of defects in the intermediate zone (zone II), as already noted, was twins of deformation origin. In the central zone I, in addition to twins, there were observed regions of $\alpha \rightarrow \varepsilon \rightarrow \alpha$ transformation and regions of localized plastic flow with a complex nonuniform structure, macro- and microcracks (indicated by arrows in Figure 7.14a). This is due to the fact that the conditions of deformation of the metal in zones I and II are significantly different. In the vicinity of the crater, higher temperatures and pressures are achieved. No signs of twinning were found in these microvolumes. The formation of zones of localized flow is a consequence of the inhomogeneous plastic flow of a material that is deformed at a high rate. Intensive heating of the central zones of the bands leads to the development of $\alpha \rightarrow \gamma \rightarrow \alpha$ transformation and formation of new equiaxed grains.

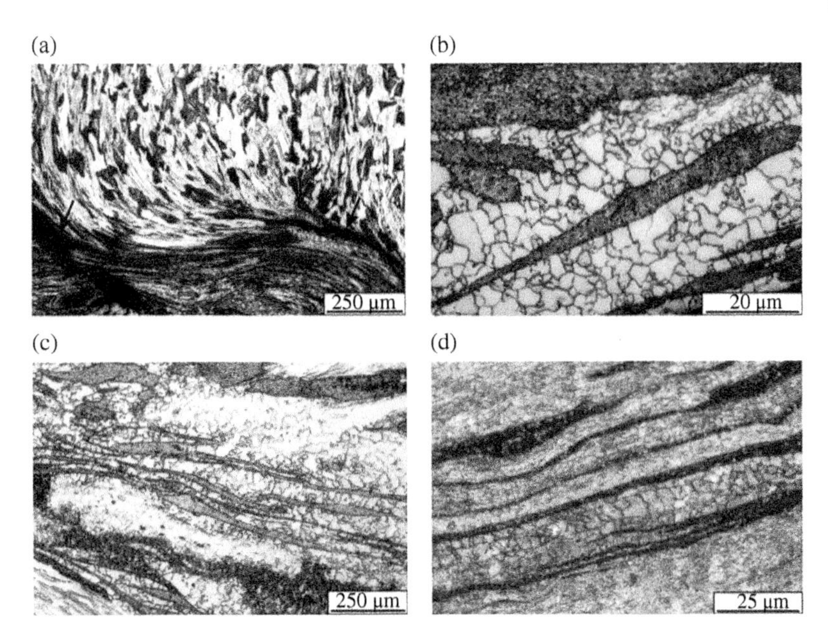

Figure 7.14 Structure of the central zone of steel disk (zone I) after the deformation using energy focusing scheme. (a–d) Show localization of plastic deformation.

The behavior of the material in the zone of localization of plastic flow resembles the features of systems with positive feedback. The acceleration of the deviation from the steady state after initiation of the process of external disturbance is typical for them. Thus, if in a system with a positive feedback the process starts for some reason, then its development is facilitated, and the system quickly leaves the steady state. In this case, we are talking about the development of the process of plastic flow of a material under dynamic loading conditions. The flow of material begins in localized zones, on a limited number of planes, and is accompanied by intense heat generation. The resulting increase in temperature leads to a decrease in the strength properties of the material in the zone of intense slip. The decrease in strength properties, in turn, facilitates the process of plastic flow and causes a further increase in temperature. Consequently, the system is moving further away from equilibrium conditions. The zones of localized plastic flow that emerged first have the highest chances for active development. The possibilities for the development of slip processes in neighboring microvolumes decrease sharply. Thus, under conditions of high-strain-rate loading, contributing to the localization of plastic flow, the resource of plasticity of the main part of the material is not consumed before the formation of cracks leading to failure.

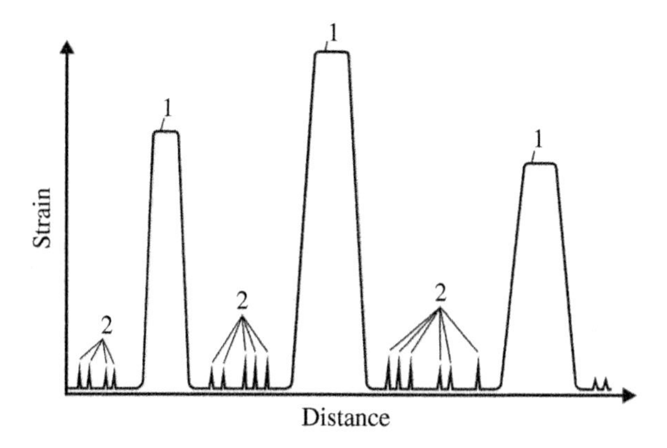

Figure 7.15 Localization of plastic deformation in a disk of low carbon steel: 1, bands of localized plastic flow, and 2, deformation twins.

Schematically, the processes of localization of plastic flow that occur under dynamic loading are shown in Figure 7.15. From this scheme, built on the basis of the model of dislocation flow instability by Vladimirov [42], it follows that the deformation processes are concentrated in narrow twins bounded by grain sizes, as well as in extended bands of localized flow that have a superstructural nature and cross multiple grains.

The amount of heat released during shear of the material along the localized flow bands is sufficient to heat the steel to temperatures corresponding to the austenitic region. The localization of the processes of plastic flow, which occurs during high strain rate loading, leads to the fact that heat is generated in a limited number of microvolumes of material in a very short time. The remaining volume of the material at the stage of its shearing along narrow bands remains in the unloaded state. Thus, a sharp temperature gradient appears in the vicinity of localized flow bands: from the melting temperature in the central part of the band to temperatures close to room temperature (at a distance of about 100–200 μm from the band). After the stage of intense shear, the temperature starts to decrease. Heat from the sheared zones goes into the volumes that remained cold. The nature of the structural transformations occurring in this process is determined by the geometrical parameters of the zones overheated during the deformation, by the rate of the heat transfer and by the distance from the overheated zones.

In the microvolumes that are in the austenitic state containing a lot of carbon, a diffusionless transformation mechanism is realized, and fine-crystalline martensite structure is formed. As a result, the central zones of the shear bands acquire an increased level of hardness. The maximum level of microhardness in the hardened zones of intensely deformed

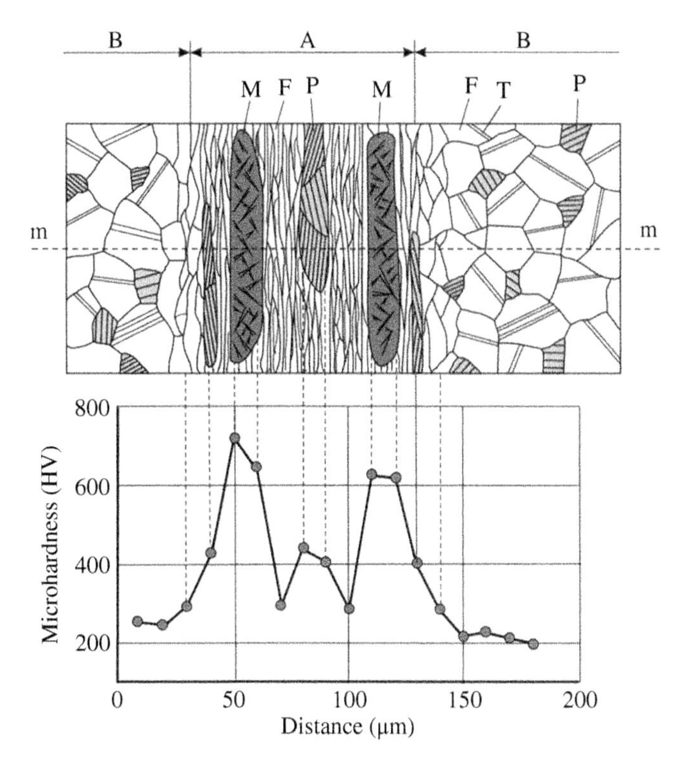

Figure 7.16 Microhardness of low carbon steel in a band of localized plastic flow: A, zone of localized plastic flow; B, zone of slightly deformed material; M, martensite; F, ferrite; P, pearlite; T, twin in ferrite; and m-m, path of microhardness measurements.

bands reaches 720 HV (Figure 7.16). The fluctuations in the properties of a material observed during the measurement of microhardness are due to the short duration of the heating process and the heterogeneity in carbon content of closely located microvolumes. The microhardness of the deformed perlite in the bands of localized deformation was 440 HV. This is 120 HV more than the microhardness of perlite in the original state (320 HV). Similar behavior was also observed for ferrite. In the initial state, the microhardness of the α-phase grains was equal to 190 HV. The severely deformed ferrite, which was located in the bands of a localized flow, was hardened to a value of 300 HV (Figure 7.16).

The heat transferred in adjacent zones when cooling overheated bands of localized plastic flow leads to structural transformations. In particular, the development of recrystallization was observed metallographically. The degree of grain refinement due to recrystallization depends on the amount of heat removed from the overheated zones, i.e. on their size and temperature of heating. The minimum size of recrystallized ferrite was in the range of 1–2 μm. It was established experimentally that in the vicinity of broad (~500–1000 μm) zones that were in the molten state, the size of recrystallized ferrite grains increases to 8–10 μm (Figure 7.14b). Large grains are also observed in the zones located between several overheated deformation bands.

The width of the bands of localized plastic flow varies significantly – from tens to hundreds of micrometers (Figure 7.14c and d). The bands have a complex structure and clearly stand out against the background of the undeformed material. Ferrite grains and pearlitic colonies, distinguishable within the bands, are elongated in the direction of shear (Figure 7.15d). The ratio of the length of deformed pearlitic colonies to their thickness reaches 15–20 or more. Ferrite grains are lengthened even more. The thickness of the individual ferrite grains in the zones of severe plastic deformation was 0.8–1.0 μm. A lot of cracks and pores of irregular shape were observed inside the bands. Their formation was due to the strong strain hardening of ferrite and perlite and the formation of brittle interlayers of fine-crystalline martensite. In zones of increased concentration of localized deformation bands, the cracks merged, and the material chipping was observed. Such a mechanism of failure is implemented mainly near the immediate vicinity of the crater.

7.4 Effect of High-Velocity Cumulative Jets on Structure of Metallic Substrates

In Chapter 5 we described phenomena, which take place during collapse of ceramic tubes by detonation products of an explosive. The resulting cumulative jets had extremely high velocity. The pressure at the collision point of the collapsing walls of the cylinder was 0.9–1.0 Mbar, the velocity of the leading part of the cumulative jet was close to 23 km/s, and the velocity of its main part was about 14 km/s. Data on the characteristics of the behavior of the material in its interaction with cumulative jets at such high speeds today is extremely small. From our point of view, of particular interest are the details of the structural transformations that take place in the course of such collisions. Currently there exists only a limited amount of data on the behavior of materials during its interaction with cumulative jets at such high velocities. For materials science, the details of the structural transformations that take place in the course of such collisions are of particular interest. This section describes the results of metallographic studies of steel plates used as witness samples in experiments on cumulative jets forming during explosive compression of ceramic tubes.

High-carbon steel containing 0.7% wt. C in a normalized condition was used as material for witness samples. Figures 5.21 and 5.22 showed the photographs of steel witness samples after collision with a cumulative jet. An interesting feature of the targets affected by the flow of ceramic particles was the abnormally large sizes (diameter and depth) of the cavities that emerged. On the front side of the samples, the diameter of the cavities was 20–25 mm. The depth of the cavities increased with increasing thickness of the ceramic tube. When using a charge with a ceramic tube with a thickness of 1.5 mm, a through penetration of the sample to a depth of 30 mm was observed.

A cross section of the sample is schematically shown in Figure 7.17. In the central part of the sample, there was a deep cavity. The walls of the cavity were covered with foreign particles (Figure 7.18a and b). The particles consisted mainly of aluminum oxides, but there were also inclusions of iron oxides. The most significant structural transformations occurred near the surface of the cavity. Figure 7.19 indicates melting of a thin surface layer due to heating of the material to a high temperature.

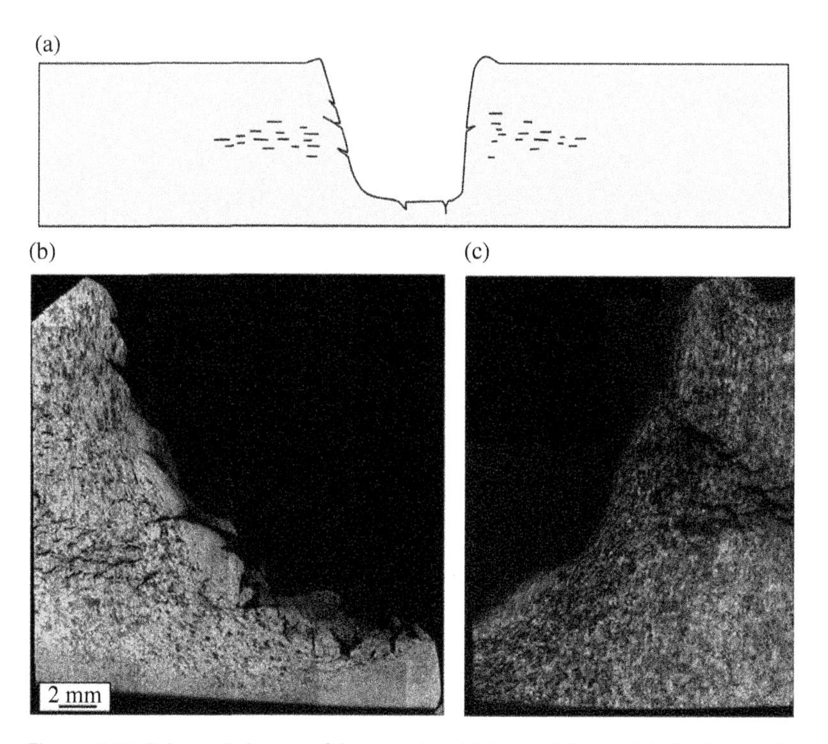

Figure 7.17 Schematic image of the sample, which was deformed during interaction with high-velocity cumulative jet (a) and corresponding images obtained by optical microscopy (b and c).

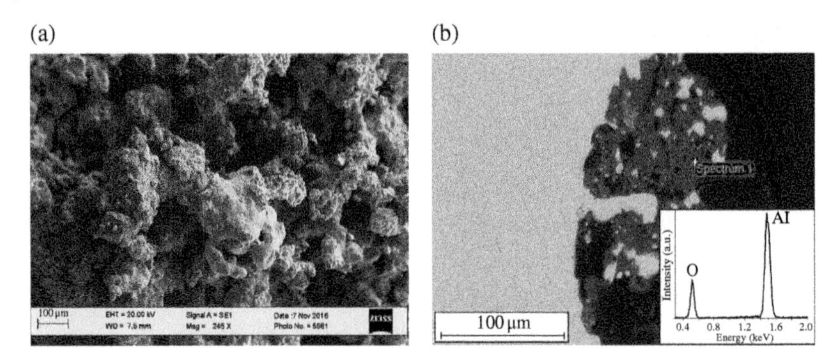

Figure 7.18 Oxide particles, which were found on the walls of the cavity by scanning electron microscope (SEM) (a) and their cross-sectional image (b).

A deeper layer of material with a thickness of about approximately 50 μm was subjected to significant heating and transition to the austenitic state. The subsequent rapid cooling, which occurred due to the removal of heat into less deformed and, accordingly, less heated surrounding volumes of the sample, led to the formation of a quenching structure consisting of martensite and residual austenite (Figure 7.19c).

The arrow in Figure 7.19c indicates a colony of perlite having plates of cementite and ferrite located almost perpendicular to the wall of the cavity. Analysis of this image allows us to conclude that the temperature exposure was short and the temperature gradient near the cavity surface was high. For instance, at a distance of more than 50 μm from the surface, the cementite plates remained practically undissolved. As the temperature approached the surface, the degree of dissolution of the cementite plates increased, and at a distance of about 15 μm from the surface the temperature was sufficient to completely dissolve them in austenite.

The cracks that appeared on the surface of the cavity spread into the adjacent volumes of material to a depth of several millimeters. From the images presented in Figure 7.19d and e, it follows that the temperature near the walls of the cracks and near their tips was abnormally high. The formation of spherical pores indicates that the material has probably reached a boiling point. It is known that the boiling point of pure iron is about 2750 °C; however, with the addition of alloying elements typical of steel, the boiling point of the alloy may decrease. Nevertheless, the presented data indicate that the growth of cracks on the surface of the cavity occurred at an extremely high rate under conditions close to adiabatic. The subsequent removal of heat to the neighboring areas led to quenching and the formation of a martensitic–austenitic structure along the cracks.

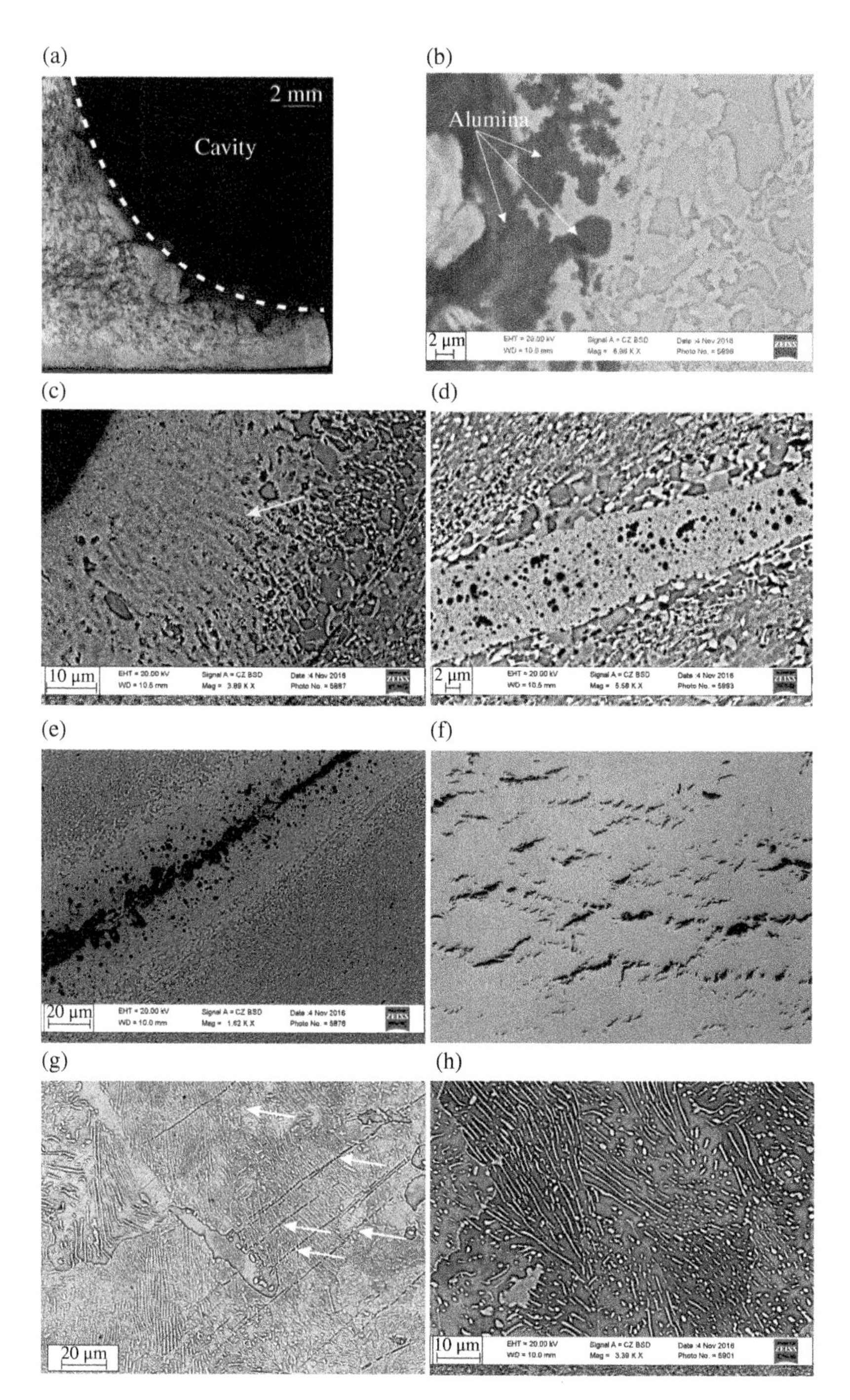

Figure 7.19 Microstructure of the sample exposed to a cumulative jet, formed during explosive compression of ceramic tubes. (a) Light microscopy and (b–h) scanning electron microscopy.

Layers of material near the crack with a width of 2–20 μm were also subjected to significant thermal impact. The result of this impact was the recrystallization of deformed ferritic grains and the quenching of micro-volumes at the site of the former pearlite colonies (Figure 7.19d).

Approximately in the middle of the sample thickness, in the zone shown in Figure 7.19f, cracking of the material was observed due to interference of the tensile waves. The absence of recrystallization in ferrite, as well as absence of phase transformations at the place of pearlite colonies, suggests that the formation of the cracks observed in Figure 7.19f occurred without a significant increase in temperature.

The propagation of shock waves in a sample is evidenced by the formation in the volume of ferrite grains and perlite colonies of multiple deformation twins (Figure 7.19g, shown by arrows). The formation of this type of defects in steels with a ferritic structure occurs under conditions of high strain rate deformation at relatively low temperatures. It should be noted that according to the hypothesis proposed in [43], deformation twins in steel arise under conditions of compressive stresses. According to Bowden and Kelly, all the observed deformation twins in pearlite were formed only in the compressed areas of the sample [43]. The diameter of the zone around the center of the sample, within which the formation of twins was observed, was approximately 45 mm. In the volumes outside of this zone, no noticeable changes in the structure were observed by SEM (Figure 7.19h).

From the experimentally obtained data, it follows that the most significant structural changes occurred near the formed cavity, which is a quite intuitive result. Similar conclusions can be drawn based on the microhardness measurements of the studied materials. The distribution

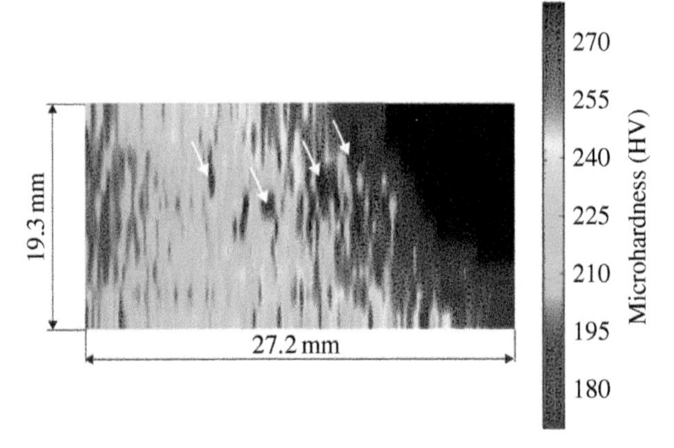

Figure 7.20 Distribution of the microhardness in the vicinity of the cavity formed by cumulative jet. The resolution of the image is 10 × 137 points. The arrows show the cracks.

of microhardness over the cross section of the sample is shown in Figure 7.20. The most significant hardening is observed in the immediate vicinity of the cavity walls. It should be noted that the depth of the hardened layer exceeds 10 mm, which is in good agreement with the data of Deribas on the hardening of steel by shock waves [44].

7.5 Summary

The results presented in this section prove that metallographic analysis of witness samples is a useful addition in experiments related to explosive technologies. Analyzing the structures of the samples, one can indirectly estimate the level of the pressures and temperatures and make a conclusion about the strain and strain rates. The results can be used to evaluate the validity of numerical simulations and increase confidence in the correctness of the calculated values. An ordinary carbon steel may be considered as a convenient material for such purposes. The advantages of carbon steel are low price, existence of phase transformations due to variations of pressure and temperature and ability to form martensite phase upon fast cooling.

References

1 Field, J.E., Walley, S.M., Proud, W.G. et al. (2004). Review of experimental techniques for high rate deformation and shock studies. *International Journal of Impact Engineering* 30 (7): 725–775.
2 Buijs, N.W. (2010). Explosive welding of metals in a vacuum environment. *Stainless Steel World* 3: 1–4.
3 Meyers, M.A. (1994). *Dynamic Behavior of Materials*. New York: John Wiley & Sons.
4 Zukas, J.A. and Walters, W.P. (2002). *Explosive Effects and Applications*. New York: Springer.
5 Selivanov, V.V., Kobylkin, I.F., and Novikov, S.A. (2014). *Explosive Technologies*. Moscow: Bauman State Technical University [in Russian].
6 Cooper, P.W. (2018). *Explosives Engineering*. John Wiley & Sons.
7 Tushinskij, L.I. (2004). *Structural Theory of Constructive Strength*. Novosibirsk: NSTU [in Russian].
8 Mecking, H. and Kocks, U.F. (1981). Kinetics of flow and strain-hardening. *Acta Metallurgica* 29 (11): 1865–1875.
9 Rusty Gray, G.T. III (2012). High-strain-rate deformation: Mechanical behavior and deformation substructures induced. *Annual Review of Materials Research* 42: 285–303.

10 Edington, J.W. (1969). The distribution of dislocations in specimens of niobium and copper after deformation in the Hopkinson bar. *Transactions of the Metallurgical Society of AIME* 245: 1653–1664.

11 Gil Sevillano, J., van Houtte, P., and Aernoudt, E. (1980). Large strain work hardening and textures. *Progress in Materials Science* 25 (2–4): 69–134.

12 Meyers, M.A. and Murr, L.E. (1981). Defect generation in shock-wave deformation. In: *Shock Waves and High-Strain-Rate Phenomena in Metals* (ed. M.A. Meyers and L.E. Murr), 487–530. Boston MA: Springer.

13 Smith, C.S. (1958). Metallographic studies of metals after explosive shock. *Transactions of the Metallurgical Society of AIME* 212: 574–589.

14 Hornbogen, E. (1962). Shock-induced dislocations. *Acta Metallurgica* 10 (10): 978–980.

15 Meyers, M.A. (1978). A mechanism for dislocation generation in shock-wave deformation. *Scripta Metallurgica* 12 (1): 21–26.

16 Meyers, M.A. (1979). A model for dislocation generation in shock-wave deformation. In: *Strength of Metals and Alloys* (ed. P. Haasen, V. Gerold and G. Kostorz), 547–552. New York: Pergamon.

17 Armstrong, R.W. and Walley, S.M. (2008). High strain rate properties of metals and alloys. *International Materials Reviews* 53 (3): 105–128.

18 Weertman, J. (1981). Moving dislocations in a shock front). *Shock Waves and High-Strain-Rate Phenomena in Metals* (ed. M.A. Meyers and L.E. Murr), 469–486. Boston: Springer.

19 Zerilli, F.J. and Armstrong, R.W. (1997). Dislocation mechanics based analysis of material dynamics behavior: enhanced ductility, deformation twinning, shock deformation, shear instability, dynamic recovery. *Journal De Physique IV* 7: C3-637–C3-642.

20 Duvall, G.E. and Graham, R.A. (1977). Phase transitions under shock-wave loading. *Reviews of Modern Physics* 49 (3): 523–579.

21 Minshall, F.S. and Zackay, V.F. (1961). The dynamic response of iron and iron alloys to shock waves. In: *Response of Metals to High Velocity Deformation* (ed. P.G. Shewmon), 249–271. New York: Interscience Publishers.

22 Kressel, H. and Brown, N. (1967). Lattice defects in shock-deformed and cold-worked nickel. *Journal of Applied Physics* 38 (4): 1618–1625.

23 Walley, S. (2007). Shear localization: a historical overview. *Metallurgical and Materials Transactions A* 38 (11): 2629–2654.

24 Hatherly, M. (1982). Deformation at high strains. In: *Strength of Metals and Alloys* (ed. R.C. Gifkins), 1181–1195. Oxford: Pergamon.

25 Zener, C. and Hollomon, J.H. (1944). Effect of strain rate upon plastic flow of steel. *Journal of Applied Physics* 15 (1): 22–32.

26 Hargreaves, C.R. and Werner, L. (1974). *Report No. AD/A006490.* Washington, DC: National Technical Information Service.

27 Rogers, H. and Shastry, C. (1981). Material factors in adiabatic shearing in steels. In: *Shock Waves and High-Strain-Rate Phenomena in Metals* (ed. M.A. Meyers and L.E. Murr), 285–298. New York: Plenum.

28 Rogers, H.C. (1983). Adiabatic shearing-general nature and material aspects. In: *Material Behavior Under High Stress and Ultrahigh Loading Rates* (ed. J.V. Mescall and V. Weiss), 101–118. New York: Plenum.

29 Grebe, H.A., Pak, H.R., and Meyers, M.A. (1985). Adiabatic shear localization in titanium and Ti-6 pct Al-4 pct V alloy. *Metallurgical Transactions A* 16 (5): 761–775.

30 Timothy, S.P. and Hutchings, I.M. (1985). The structure of adiabatic shear bands in a titanium alloy. *Acta Metallurgica* 33 (4): 667–676.

31 Dodd, B. and Bai, Y. (1987). *Ductile Fracture and Ductility: With Applications to Metalworking.* London: Academic Press.

32 Grady, D.E. and Kipp, M.E. (1985). Growth of inhomogeneous thermoplastic shear. *Journal de Physique Colloques* 46: C5-291–C5-298.

33 Kipp, M.E. and Grady, D.E. (1985). Dynamic fracture growth and interaction in one dimension. *Journal of the Mechanics and Physics of Solids* 33 (4): 399–415.

34 Giovanola, J.H. (1988). Adiabatic shear banding under pure shear loading part i: direct observation of strain localization and energy dissipation measurements. *Mechanics of Materials* 7 (1): 59–71.

35 Marchand, A. and Duffy, J. (1988). An experimental study of the formation process of adiabatic shear bands in a structural steel. *Journal of the Mechanics and Physics of Solids* 36 (3): 251–283.

36 Gotoh, M., Yamashita, M., and Ohno, M. (1992). An evidence of melting along adiabatic shear band in high-speed shearing process. *Transactions of the Japan Society of Mechanical Engineers Series A* 58 (554): 1979–1984.

37 Russell, A., Schmelzer, J., Müller, P. et al. (2015). Mechanical properties and failure probability of compact agglomerates. *Powder Technology* 286: 546–556.

38 Mei, Z.G., Shang, S.L., Wang, Y., and Liu, Z.K. (2009). Density-functional study of the thermodynamic properties and the pressure-temperature phase diagram of Ti. *Physical Review B - Condensed Matter and Materials Physics* 80 (10): 104116.

39 Dinsdale, A.T. (1991). SGTE data for pure elements. *Calphad* 15 (4): 317–425.

40 Kheifets, A.E., Zel'dovich, V.I., Frolova, N.Y. et al. (2017). Phase and structural transformations in a low-carbon steel that occur upon the collapse of a cylindrical shell. *Physics of Metals and Metallography* 118 (7): 681–690.

41 Goldshtein, M.I., Litvinov, V.S., and Bronfin, B.M. (1986). *Physics of Highstrength Alloys*. Moscow: Metallurgy [in Russian].
42 Vladimirov, V.I. (1984). *Physical Nature of Failure of Metals*. Moscow: Metallurgy [in Russian].
43 Bowden, H.G. and Kelly, P.M. (1967). Deformation twinning in shock-loaded pearlite. *Acta Metallurgica* 15 (1): 105–111.
44 Deribas, A.A. (1980). *Physics of Explosive Hardening and Welding*. Novosibirsk: Nauka [in Russian].

Conclusions

Out of many processes occurring in explosive substances, detonation is of a special interest. Development of explosive devices and technologies usually requires estimates and calculations that are most often based on the hydrodynamic theory (model) of detonation developed in the works of Zeldovich, Neumann, and Döring. The initial assumptions of the model are formulated for a plane detonation wave, which obviously does not correspond to the detonation in real charges that always have finite dimensions. This leads to the concept of a limiting and a critical diameter (thickness) of the charge. In the scientific community, a discussion on the degree of adequacy and the boundaries of applicability of the hydrodynamic model is still ongoing. In the experiments, several effects were discovered that do not fit into the framework of this model.

For example, data revealing the development of chemical processes behind the Chapman–Jouguet plane (behind the postulated reaction zone) are under discussion. Disturbances in the stationarity of the process are also found when using charges of complex geometric shapes, in experiments when the detonation front comes out into the expanding region (which leads to the formation of areas with unreacted explosives), and in the presence of cavities or gaps between charges and inert elements and so on.

A special attention should be paid to detonation processes in charges contacting with inert elements of materials having a sound velocity exceeding the detonation velocity. In such cases, shock waves excited by detonation in inert materials can outrun the front of the detonation wave and compress the explosive substance ahead of the front. This leads to a change in the state of explosives and a corresponding change in the kinetics of the detonation transformation. As a result, the stationarity of the process is disrupted, which can lead to hardly predictable changes in

Explosion Systems with Inert High-Modulus Components: Increasing the Efficiency of Blast Technologies and Their Applications, First Edition. Igor A. Balagansky, Anatoliy A. Bataev, and Ivan A. Bataev.

its course and, from a practical point of view, to a decrease or an increase in the effectiveness of a particular explosive technology. On the other hand, an understanding of the mechanisms for the generation, development, and realization of nonstationary effects in detonation can provide a tool to increase the effectiveness of explosive technologies and/or purposeful control of detonation processes.

The monograph summarizes the results of experimental, numerical, and theoretical studies of explosion systems that contain high-modulus ceramic elements. The phenomena arising in such systems are described in detail: desensitization of explosives, nonstationary detonation processes, the cumulation of energy, and the emergence of Mach waves. The formation of high-speed flows of ceramic particles arising during the explosive collapse of ceramic tubes is described. Structural transformations in metals due to the focusing of the explosion energy or the action of a superfast ceramic jet are discussed. These transformations include, but are not limited to, adiabatic shear banding, phase transformations, twinning, melting, boiling, and even evaporation of impacted substrates.

Taking into account the described phenomena and having an aim to increase the effectiveness of explosive technologies, the following should be noted:

1) The phenomenon of desensitization of explosives by an advanced compression wave can be used to create explosion-proof structures based on high-modulus ceramic materials.
2) The occurrence of nonstationary detonation regimes should be taken into account when designing explosion devices containing materials with a high sound velocity.
3) When choosing explosion welding modes, the potential influence of advanced waves in the projected plate on the processes in the welding zone should be evaluated.
4) Mach detonation waves emerging under appropriate conditions can significantly improve the parameters of explosive loading of materials.
5) Energy cumulation processes in the explosive collapse of ceramic tubes make it possible to obtain particle flows at a rate exceeding the threshold velocity at which melting and partial evaporation of the target material occurs.
6) When metals are loaded, specific conditions may be achieved, which make it possible to obtain structures possessing properties useful for practical applications.

The authors hope that the results presented in this monograph will be useful for modernizing and enhancing the effectiveness of both existing and future explosion technologies.

List of the Main Publications of Authors on the Theme of Monograph

1 Balagansky, I.A., Kobylkin, I.F., Razorenov, S.V. et al. (1991). Effect of a silicon carbide shell on detonation parameters in high explosives. *Proceedings of 5 All-Union Conference on Detonation.* Krasnoyarsk, USSR (5–12 August 1991). Krasnoyarsk: Russian Academy of Science. [in Russian].

2 Balagansky, I.A., Berdnik, V.P., Kulikova, I.V. et al. (1991). Features of detonation processes in HE charges that contact with high modulus ceramics. *Annotations of the reports of the 7th All-Union Congress on Theoretical and Applied Mechanics.* Moscow, USSR (15–21 August 1991). Moscow: Moscow State University. [in Russian].

3 Balagansky, I.A., Razorenov, S.V., and Utkin, A.V. (1993). Detonation parameters of condensed high explosive charges with long ceramic elements. *Proceedings of the 10th International Detonation Symposium,* Boston, USA (12–16 July 1993). Arlington: Office of Naval Research.

4 Balaganskii, I.A., Agureikin, V.A., Razorenov, S.V. et al. (1994). Effect of an inert high-modulus ceramic wall on detonation propagation in solid explosive charges. *Combustion. Explosion and Shock Waves* 30: 674–681.

5 Balagansky, I.A. and Gryaznov, E.F. (1994). Desensitization of RDX-charges after preshocking by compression wave in SiC-ceramic rod. *Proceedings of International Conference on Combustion 'Zel'dovich Memorial',* Moscow, Russia (12–17 September 1994). Moscow: Russian Section of the Combustion Institute.

6 Balagansky, I.A., Zorin, S.V., and Kaplouhov, V.M. (1995). Container for transportation and storage of explosive materials. Russian Federation Patent 2094751 filed 11 January 1995 and issued 27 October 1997.

7 Balagansky, I.A., Agureikin, V.A., Kobilkin, I.F. et al. (1999). Acceleration device based on high explosive charge, which contains high modular ceramic tube. *International Journal of Impact Engineering* 22: 813–823.

8 Balagansky, I.A., Vinogradov, A.V., Agureikin, V.A. et al. (2002). Explosion systems based on HE explosion charges, which include high-modulus ceramic tubes. *Proceedings of the International Conference 'III Khariton Thematic Scientific Readings',* Sarov, Russia (26 February–2 March 2001). Sarov: RFNC. [in Russian].

9 Balagansky, I.A., Balagansky, A.I., Razorenov, S.V. et al. (2005). Evolution of shock waves in silicon carbide rods. *Proceedings of the International Conference 'VII Khariton Topical Scientific Readings',* Sarov, Russia (14–18 March 2005). Sarov: RFNC. [in Russian].

10 Balagansky, I.A., Balagansky, A.I., Kobilkin, I.F. et al. (2005). Influence of high explosive charge shell on detonation front shape. *Proceedings of International conference 'VIII Zababakhin Scientific readings',* Snezhinsk, Russia (5–9 September 2005). Snezhinsk: RFNC. [in Russian].

11 Balagansky, I.A., Balagansky, A.I., Razorenov, S.V. et al. (2006). Evolution of shock waves in silicon carbide rods. *Proceedings of the 14th APS Topical Conference on Shock Compression of Condensed Matter,* Baltimore, USA (31 July–5 August 2005). Melville: AIP Publishing.

12 Balagansky, I.A., Matrosov, A.D., Stadnichenko, I.A. et al. (2007). Influence of inert copper and silicon carbide inserts on process of detonation transmission through water. *Proceedings of the International Conference 'IX Khariton Topical Scientific Readings',* Sarov, Russia (12–16 March 2007). Sarov: RFNC. [in Russian].

13 Balagansky, I.A., Matrosov, A.D., Stadnichenko, I.A. et al. (2008). Influence of inert copper and silicon carbide inserts on process of detonation transmission through water. *Materials Science Forum* 566: 207–212.

14 Balagansky, I.A., Matrosov, A.D., Stadnichenko, I.A. et al. (2008). Desensitization of heterogeneous high explosives under initiation through high modulus elastic elements. *International Journal of Modern Physics B* 22: 1305–1310.

15 Balagansky, I.A., Matrosov, A.D., and Stadnichenko, I.A. (2008). High-modulus heterogeneous dispersive media as a material of protective shells. *Proceedings of 3 Russian conference 'Problems of explosion protection and counter-terrorism'.* Saint-Petersburg, Russia (20–21 May 2008). Saint-Petersburg: St. Petersburg State University of Emergency Situations. [in Russian].

16 Balagansky, I.A., Hokamoto, K., Manikandan, P. et al. (2009). Phenomena of energy focusing in explosive systems, which include high modulus elastic elements. *Proceedings of the 16th American Physical Society Topical Conference on Shock Compression of Condensed Matter,* Nashville, USA (28 June–3 July 2009). Melvill: American Institute of Physics.

17 Balagansky, I.A., Hokamoto, K., Manikandan, P. et al. (2010). Study of energy focusing phenomenon in explosion systems, which include high modulus elastic elements. *Proceedings of the 14th International Detonation Symposium,* Coeur d'Alene, USA (11–16 April 2010). Arlington: Office of Naval Research.

18 Balagansky, I.A., Hokamoto, K., Manikandan, P. et al. (2011). Mach stem formation in explosion systems, which include high modulus elastic elements. *Journal of Applied Physics* 110: 123516.

19 Merzhievskii, L.A., Balaganskii, I.A., Matrosov, A.D. et al. (2012). Detonation transmission through high-modulus dispersed media. *Combustion, Explosion and Shock Waves* 48: 709–712.
20 Balagansky, I.A., Vinogradov, A.V., Merzhievsky, L.A. et al. (2016). Analysis of shell material influence on detonation process in high explosive charge. *Key Engineering Materials* 715: 27–32.
21 Balagansky, I.A. and Stepanov, A.A. (2016). Numerical simulation of Composition B high explosive charge desensitization in gap test assembly after loading by precursor wave. *Shock Waves* 26: 109–115.
22 Balagansky, I.A., Vinogradov, A.V., and Merzhievsky, L.A. (2017). Modeling of fast jet formation under explosion collision of two-layer alumina/copper tubes. *The International Journal of Multiphysics* 11 (3): 265–375.
23 Balagansky, I.A., Vinogradov, A.V., Merzhievsky, L.A. et al. (2018). Analysis of the influence of the shell material on the detonation process in high explosive charge. *Combustion, Explosion and Shock Waves* 54 (4): 502–510.
24 Balaganskii, I.A., Merzhievskii, L.A., Ul'yanitskii, V.Y. et al. (2018). Generation of hypervelocity particle flows by explosive compression of ceramic tubes. *Combustion, Explosion and Shock Waves* 54: 119–124.

Appendix A

Dynamic Properties of High-Modulus Materials

A.1 Elastic Properties of Isotropic Polycrystalline Materials

A characteristic feature of elastic isotropic polycrystalline materials is that various perturbations propagate in them at different rates. The main types of disturbances (longitudinal, transverse (or shear), bulk) propagate with the corresponding velocities C_1 (longitudinal sound velocity), C_2 (shear sound velocity), and C_0 (bulk sound velocity). The waves in rods propagate with a specific speed C_s (rod sound velocity). These velocities are related to the elastic moduli by the following formulas:

$$C_1 = \sqrt{\frac{E(1-\nu)}{\rho(1+\nu)(1-2\nu)}}; \quad C_2 = \sqrt{\frac{G}{\rho}}; \quad C_0 = \sqrt{\frac{K}{\rho}}; \quad C_s = \sqrt{\frac{E}{\rho}},$$

where E is the Young's modulus, G is the shear modulus, K is the bulk compression module, ρ is the density of the material, and ν is Poisson's ratio.

The characteristics of the elastic properties, which have a practical interest for research, are given in Table A.1, according to the data of [1–12]. For the studies described in this monograph, the most interesting are high-modulus materials (mainly ceramics).

Explosion Systems with Inert High-Modulus Components: Increasing the Efficiency of Blast Technologies and Their Applications, First Edition. Igor A. Balagansky, Anatoliy A. Bataev, and Ivan A. Bataev.

A.2 Peculiarities of Wave Processes in High-Modulus Materials

The basic equation describing the behavior of continuous media in shock wave processes is the shock adiabat (the Hugoniot adiabat). A qualitative form of the shock adiabat of an elastoplastic body is shown in Figure A.1, where the specific volume V is plotted along the abscissa axis, pressure p is along the ordinate axis, the solid line corresponds to the shock adiabat and the release adiabat, the dotted line describes the hydrostatic compression curve, σ_T is the yield strength, and σ_g is the Hugoniot elastic limit (HEL). The HEL is related to the yield strength by the relationship

$$\sigma_g = \frac{1-\nu}{1-2\nu}\sigma_T = \left(\frac{K}{2G}+\frac{2}{3}\right)\sigma_T.$$

A typical feature of ceramic materials is the high values of the HEL and the velocities of the wave propagation. This determines the possibility of transmitting perturbations of a large amplitude in an elastic wave. Thus, in [13], the ability of glass and quartz to transmit large pressures in an elastic wave over considerable distances is described. In [14], the behavior of a ceramic material based on titanium carbide (80% by mass) under shock wave loading was studied. The authors observed an increase in the particle velocity dispersion at the compression wave front while increasing the sample thickness. Another important feature of ceramic materials is that under shock wave loading, intensive failure due to the action of tangential stresses occurs, while plastic deformation is practically absent. Figure A.2 shows a typical diagram of states for brittle materials [15].

The values of the HEL for several ceramic materials are given in Table A.2.

Under normal conditions, the cracking of ceramics commences under compressive stresses at 0.33–0.66 of the elastic limit [16]. With a further increase in shearing stresses, cracks start growing and merging; the material breaks down and then behaves like a powder or sand.

Different types of ceramic materials differ in the behavior of deformation during shock wave compression [3]. In some cases, as the load increases above the HEL, the state of the shock-compressed material approaches the hydrostatic one. Investigations of samples of single-crystal quartz preserved after the shock wave action showed that they consist of crystalline blocks separated by interlayers of quartz glass, which is a consequence of the development of adiabatic shears under dynamic loading. Thus, there exists enough evidence of a heterogeneous behavior under the dynamic deformation of high strength materials.

Table A.1 Characteristics of the elastic properties.

Material	E, GPa	G, GPa	K, GPa	N	ρ_0, g/cm^3	C_1, km/s	C_2, km/s	C_0, km/s	C_s, km/s
Be	293	131	127	0.115	1.82	12.9	8.5	8.4	12.7
Li	11.8	4.2	14.3	0.336	0.54	5.8	2.8	5.2	4.7
Si	171	71	100	0.214	2.34	9.1	5.5	6.5	8.6
BN	810	330	422	0.180	2.30	19.6	12.0	13.5	18.8
B_4C	474	201	247	0.180	2.52	14.3	8.9	9.9	13.7
Al_2O_3	403	163	249	0.230	3.96	10.9	6.4	7.9	10.1
BeO	394	159	250	0.234	3.01	12.4	7.3	9.1	11.4
SiC	401	170	234	0.170	3.22	11.6	7.3	8.5	11.2
AlB_{12}	352	163	139	0.077	2.54	11.8	8.0	7.4	11.7
WC	700	280	620	0.310	15.01	8.0	4.3	6.4	6.8
SiO_2 melted	96	44	38	0.078	2.65	6.1	4.1	3.8	6.0
SiO_2 crystalline	101	42	56	0.203	2.77	6.4	3.9	4.5	6.0

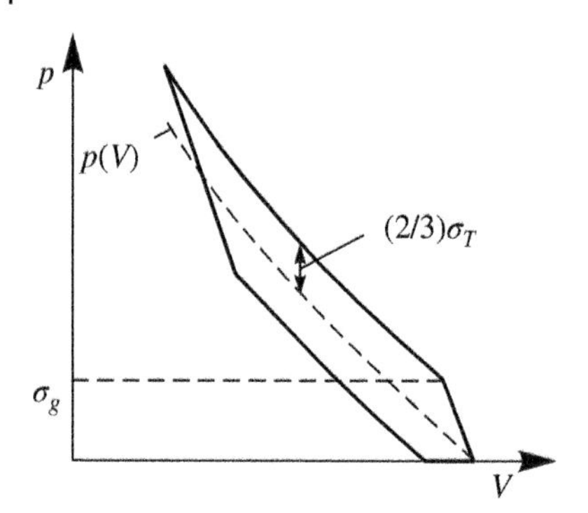

Figure A.1 Shock adiabat of an elastoplastic body.

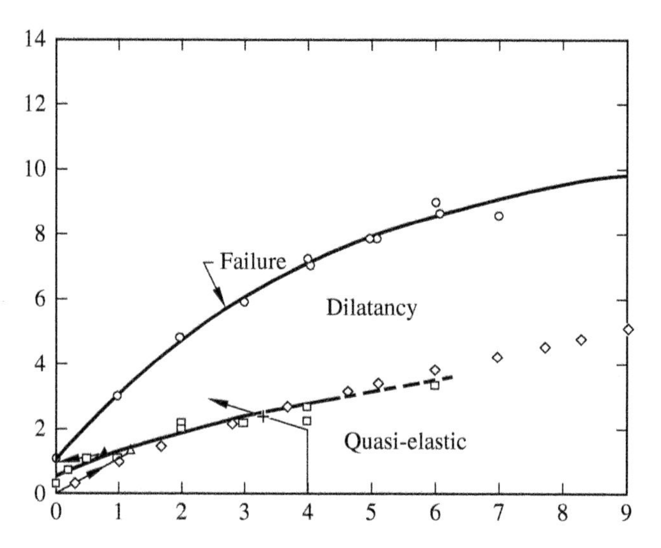

Figure A.2 The failure envelope for dry granodiorite (shear stress vs. pressure, kbar). (*Source:* From Schock et al. [15]. Reprinted with permission of John Wiley & Sons.)

Table A.2 The values of the Hugoniot elastic limit (HEL).

Material	B_4C	Al_2O_3	BeO	SiC	AlB_{12}	WC	SiO_2 melted	SiO_2 crystalline
σ_g (HEL)	15.4	11.2	8.2	8–15	8.7	4.6	9.8–26.0	14.4–38.3

With impact compression of ceramic materials, polymorphic transformations often occur, which affects the distribution of compression and rarefaction waves. Current studies indicate that phase transitions in silicate rocks lead to stronger attenuation of amplitude parameters of shock waves and to a qualitative change in the velocity fields, and the fraction of energy dissipated in the phase transition zone constitutes a considerable part of the total energy.

A.3 Characteristics of Ceramic Materials Used in the Experiments

In the experiments described in Chapters 2–6, the samples from solid and powdered silicon carbide SiC, aluminum oxide (alumina) Al_2O_3, and silica (river sand) SiO_2 were used.

Around 250 crystalline forms of silicon carbide are known. The polymorphism of silicon carbide is characterized by a large number of similar crystalline structures called polytypes. They are variations of the same chemical compound, which are identical in two dimensions but differ in the third. Thus, they can be considered layers stacked in a certain sequence [17]. Alpha-silicon carbide (α-SiC) is the most common polymorph. This modification is formed at the temperature above 1700 °C and has a hexagonal lattice, a wurtzite-type crystal structure.

Beta modification (β-SiC) with a crystal structure of the zinc blende type (similar to the structure of diamonds) is formed at temperatures below 1700 °C. Heating the beta modification of SiC to temperatures above 1700 °C can lead to a gradual transition of the cubic lattice to hexagonal (2H, 4H, 6H, 8H) and rhombic (15R). With an increase in the temperature and time of the heating process, all forms of SiC eventually transform to the hexagonal alpha polytype 6H. The structures of the basic SiC polytypes are shown in Figure A.3. Table A.3 shows the corresponding types of crystal structures.

Such a huge variety of crystal structures is reflected in the shock adiabat (the dependence of pressure on the ratio of the initial and final specific volumes), and Figure A.4 shows the shock adiabat and some specific points (the HEL, the points of transition to various polytypes). The dotted lines correspond to hydrostatic compression curves [18].

Alumina also has several modifications, the main ones of which are given in Table A.4 [19].

In nature, one can find only the trigonal α-modification of aluminum oxide as a mineral called corundum and its rare precious varieties (ruby, sapphire, etc.). It is the only thermodynamically stable form of Al_2O_3. Heat treatment of aluminum hydroxides at about 399 °C leads to the formation of a cubic γ-modification. At 1100–1200 °C the γ-modification undergoes an irreversible transformation to α-Al_2O_3. However, the speed of this process is rather slow.

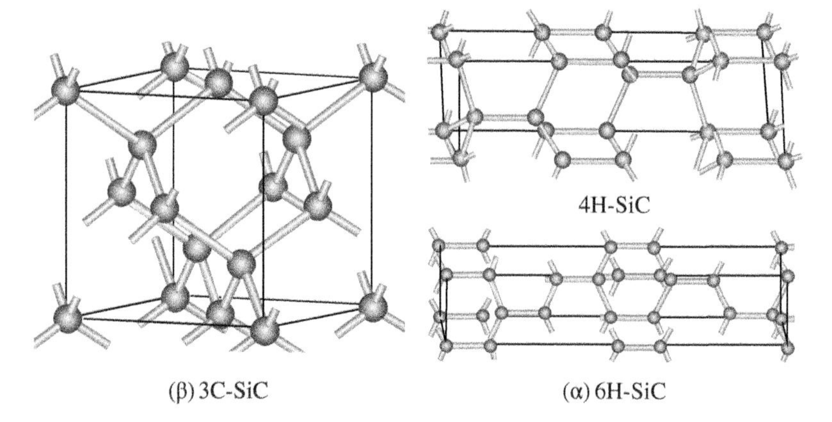

Figure A.3 Types of crystal lattices of silicon carbide (https://en.wikipedia.org/wiki/Silicon_carbide).

Table A.3 Types of crystal lattices of silicon carbide.

Polytype	3C (β)	4H	6H (α)
Crystal structure	Cubic (zinc blende)	Hexagonal	Hexagonal

Thus, for the complete phase transition, it is necessary either to have stabilizing elements or to increase the processing temperature to 1400–1450 °C. The following crystalline modifications of aluminum oxide are also known: a cubic η-phase, a monoclinic θ-phase, a hexagonal χ-phase, and an orthorhombic κ-phase. The existence of the δ-phase, which can be tetragonal or orthorhombic, remains a questionable topic.

Shock adiabats of several commercially available types of alumina are shown in Figure A.5 [20].

In [21], to build a mathematical model of the dynamic compression of polycrystalline Al_2O_3, Maxwellian ideas about the mechanisms of irreversible deformation are involved. The calculations made it possible to trace the changes in the tangential stress (Y) behind the front of the shock wave depending on its amplitude σ_1 (Figure A.6; here σ_H is HEL, solid line corresponds to calculation, and diamond points correspond to the experimentally obtained data).

Experimental shock adiabats of silica (river sand) were obtained in [22–25].

An example of such shock adiabats for sand of various fractions is shown in Figure A.7, where the designations of the curves correspond to those given in Table A.5, which give information about the composition

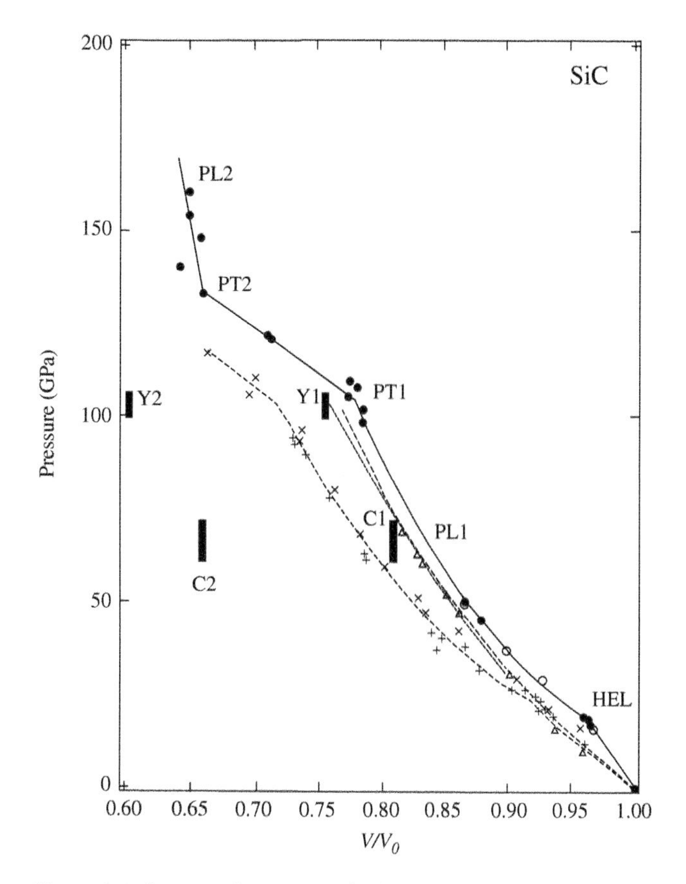

Figure A.4 Compression curves of SiC. (*Source:* From Sekine and Kobayashi [18]. Reprinted with permission of American Physical Society.)

Table A.4 Modifications of alumina.

Modification	α-Al_2O_3	θ-Al_2O_3	γ-Al_2O_3	κ-Al_2O_3
Density, g/cm^3	3.99	3.61	3.68	3.77

and humidity. The figure shows the influence of the particle size of sand and the content of water in it on its compressibility.

In the experimental studies described in this monograph, the samples shaped as square-shaped rods, plates, and rings made of self-bonded silicon carbide [26] were mainly used. The samples were manufactured at the Institute of Problems of Materials Science of the Academy of Sciences

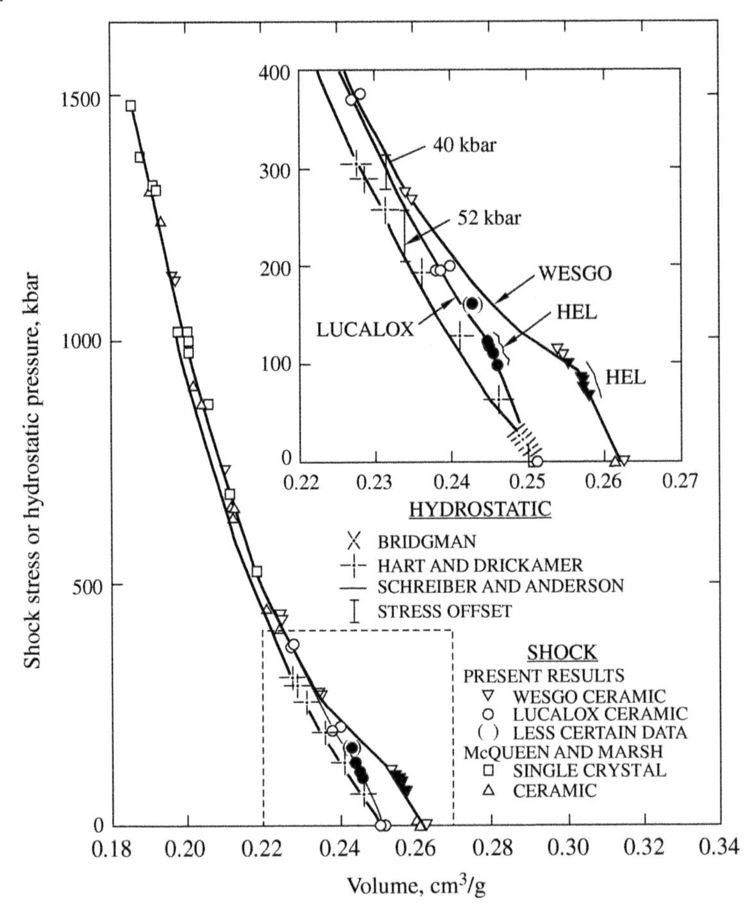

Figure A.5 Shock stress or hydrostatic pressure versus volume for various forms of aluminum oxide. Solid symbols indicate HEL states. (*Source:* From Ahrens et al. [20]. Reprinted with permission of AIP Publishing.)

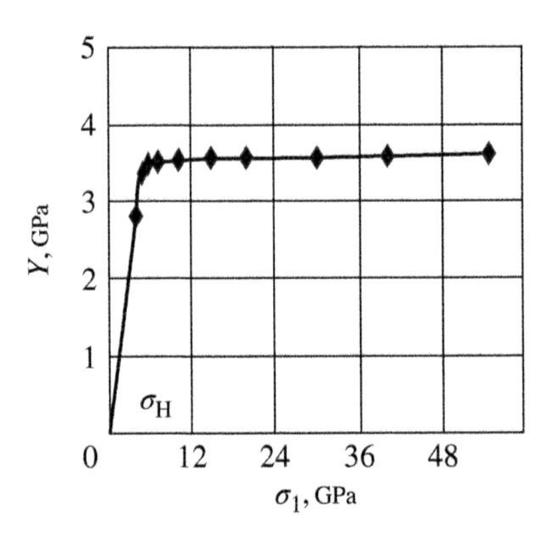

Figure A.6 Tangential stress behind the shock wave front versus its amplitude (σ_H is Hugoniot elastic limit). (*Source:* From Merzhievskii [21]. Reprinted with permission of Springer Nature.)

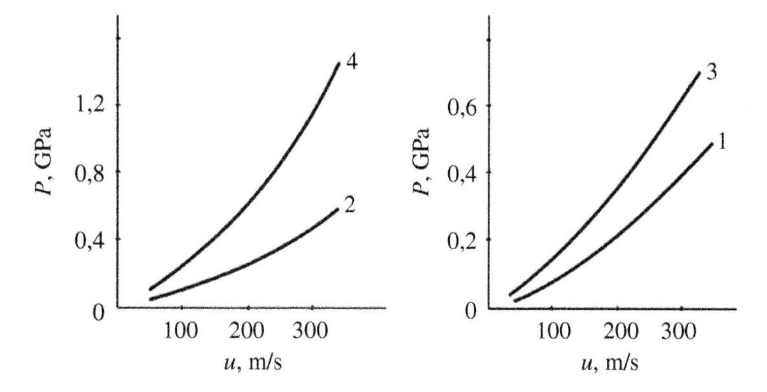

Figure A.7 Shock adiabat of river sand.

Table A.5 Composition and humidity of river sand.

Variant	Condition	Particle size, mm	Density, g/cm^3
1	Dry	0.16–0.20	1.52
2	Dry	0.20–0.315	1.6
3	Dry	0.63–1.0	1.67
4	Wet, 18 %	0.20–0.315	1.9

of Ukraine. In contrast to mass-produced materials, the samples had fewer impurities and the lowest possible porosity.

The self-bonded silicon carbide used for the samples had the following properties:

- sound velocity: 11 km/s.
- static elastic modulus: 350–360 GPa.
- ultimate static strain: 3.5–5.5 × 10^{-4}.

Densities and amounts of impurities are given in Table A.6 separately for rods, plates, and rings. The sintered samples were subjected to a diamond polishing of the ends and the inner surface. The surface of the plates that were in contact with manganin gauges was polished as well.

In the first series of experiments on the explosive collapse, the tubes of various sizes made of pure alumina produced by the Poltava plant for high-pressure gas-discharge sodium lamps were used. The proprietary name of this material is "Polikor," and it is a pure translucent corundum of density ρ_0 = 3.96 g/cm^3 with a porosity of about 0.5%. The amount of MgO impurities was less than 1%. The sound velocity was about 10 km/s.

Table A.6 Characteristics of self-bonded silicon carbide samples.

Product name, dimensions	Density ρ_0, g/cm^3	Bending strength, kgf/mm^2	Free silicon, %	Free carbon, %
Rods				
20 × 20 × 80 mm	3.08	20–22	9.5	3
63 × 60 × 10 mm	3.10	20–22	11.5	3
Rings, D = 74 mm, d = 48 mm, l = 60 mm	3.05	18–19	13.0	3

During the experiments, dispersed media based on silicon carbide (SiC) powder and river sand (silica, SiO_2) of different granulometric composition in a dry and water-filled state were also used.

References

1 Franzevich, I.N., Voronov, F.F., and Bakuta, S.A. (1982). *Elastic Constants and Elastic Moduli of Metals and Nonmetals*. Kiev: Naukova Dumka [in Russian].

2 Andrievsky, R.A. and Spivak, I.I. (1989). *Strength of Refractory Compounds and Materials Based on Them*. Chelyabinsk: Metallurgy [in Russian].

3 Bushman, A.V., Kanel, G.I., Ni, A.L. et al. (1988). *Thermophysics and Dynamics of Intensive Impulse Effects*. Chernogolovka: IPCP RAS [in Russian].

4 Ahrensf, T.J., Gust, W.H., and Royce, E.B. (1988). Material strength effect in the shock compression of alumina. *Journal of Applied Physics* 39 (10): 4610–4616.

5 Gust, W.H. and Royce, E.B. (1971). Dynamic yield strength of B_4C, BeO, and Al_2O_3 ceramics. *Journal of Applied Physics* 42 (1): 276–295.

6 Wackerly, J. (1962). Shock-wave compression of quartz. *Journal of Applied Physics* 33 (3): 922–937.

7 Grady, D.E. (1994). Hydrodynamic compressibility of silicon carbide through shock compression of metal-ceramic mixtures. *Journal of Applied Physics* 75 (1): 197–202.

8 Grady, D.E. (1994). Shock-wave strength properties of boron carbide and silicon carbide. *Journal de Physique IV* 4 (C8): 385–391.

9 Holmquist, T.J. and Johnson, G.R. (2008). Response of boron carbide subjected to high-velocity impact. *International Journal of Impact Engineering* 35: 742–752.

10 Lundberg, P., Renstrom, R., and Lundberg, B. (2006). Impact of conical tungsten projectiles on flat silicon carbide targets: transition from interface defeat to penetration. *International Journal of Impact Engineering* 32: 1842–1856.

11 Kanel, G.I., Zaretsky, E.B., Rajendran, A.M. et al. (2009). Search for conditions of compressive fracture of hard brittle ceramics at impact loading. *International Journal of Plasticity* 25 (4): 649–670.

12 Kanel, G.I., Razorenov, S.V., and Fortov, V.E. (2004). *Shock-Wave Phenomena and the Properties of Condensed Matter*. New York: Springer-Verlag.

13 Dremin, A.N. and Adadurov, G.A. (1964). Behavior of glass under dynamic loading. *Physics of the Solid State* 6 (6): 1757–1763. [in Russian].

14 Kanel, G.I. and Pityulin, A.N. (1984). Shock-wave deformation of titanium carbide-based ceramics. *Combustion, Explosion and Shock Waves* 20: 436–438.

15 Schock, R.N., Heard, H.C., and Stephens, D.R. (1973). Stress-strain behavior of a granodiorite and two greywackes sandstones on compression to 20 kilobars. *Journal of Geophysical Research* 78: 5922–5941.

16 Brace, W.F., Paulding, B.W., and Scholz, C. (1966). Dilatancy in the fracture of crystalline rocks. *Journal of Geophysical Research* 71 (16): 3939–3953.

17 Morkoç, H., Strite, S., Gao, G.B. et al. (1994). Large-band-gap SiC, III-V nitride, and II-VI ZnSe-based semiconductor device technologies. *Journal of Applied Physics* 76: 1363–1398.

18 Sekine, T. and Kobayashi, T. (1997). Shock compression of 6H polytype SiC to 160 GPa. *Physical Review B* 55 (13): 8034–8037.

19 Levin, I. and Brandon, D. (1998). Metastable alumina polymorphs: crystal structures and transition sequences. *Journal of the American Ceramic Society* 81 (8): 1995–2012.

20 Ahrens, T.J., Gust, W.H., and Royce, E.B. (1968). Material strength effect in the shock compression of alumina. *Journal of Applied Physics* 39: 4610–4616.

21 Merzhievskii, L.A. (1998). Simulation of the dynamic compression of polycrystalline Al_2O_3. *Combustion, Explosion and Shock Waves* 34: 679–687.

22 Dianov, M.D., Zlatin, N.A., Mochalov, S.M. et al. (1976). Shock compressibility of dry and water-saturated sand. *Pis'ma v Zhurnal tekhnicheskoĭ fiziki* 2 (12): 529–532. [in Russian].

23 Lagunov, V.A. and Stepanov, V.A. (1963). Measurement of dynamic compressibility of sand at high pressures. *Journal of Applied Mechanics and Technical Physics* 1: 88–96. [in Russian].

24 Bragov, A.M. and Grushevsky, G.M. (1993). Influence of humidity and granulometric composition on impact compressibility of sand. *Pis'ma v Zhurnal tekhnicheskoĭ fiziki* 19 (12): 70–72. [in Russian].

25 Grigoryan, S.S., Evterev, A.S., and Zamyshlyaev, B.V. (1978). On shock-wave processes in silicate rocks with phase transformations taken into account. *Doklady Akademii Nauk SSSR* 241 (6): 1295. [in Russian].

26 Gnesin, G.G. (1977). *Silicon Carbide Materials*. Moscow: Metallurgy [in Russian].

Appendix B

Methods Used to Investigate Explosion Systems Containing High-Modulus Inert Materials

Several traditional experimental methods for studying explosive and shock wave processes were developed simultaneously with the improvement of dynamic methods of high-pressure physics [1]. A detailed description of the entire spectrum of existing methods is given in monographs [2–3]. In current studies, we used the following techniques:

- Manganin gauges.
- Optical methods.
- The method of barriers (i.e. the GAP test).

The numerical simulation of the processes was conducted using POTOK-ES software and its later version ODVAX implemented on the MicroVAX computer [4, 5] and the ANSYS AUTODYN software package [6].

B.1 Methods of Experimental Investigations

B.1.1 Manganin Gauges

Currently, manganin gauges are one of the most popular methods to continuously record values of pressures (stresses) inside samples or on contact and free surfaces under dynamic loading. This method was first used to measure pressure under static loads [7]. Further development of the method can be found, for example, in [2–3, 8–13].

In most of the experiments described in the monograph, the gauges had the shape of a sinusoid, as shown in Figure B.1. They are made of manganin foil with a thickness of 20 μm. The width of the sinusoidal line is approximately 0.1–0.2 mm. The accuracy of measurements over time is not worse than 0.01 μs. The accuracy of the pressure measurement is 3–5%. The temporal resolution is determined primarily by the thickness of the insulation,

Explosion Systems with Inert High-Modulus Components: Increasing the Efficiency of Blast Technologies and Their Applications, First Edition. Igor A. Balagansky, Anatoliy A. Bataev, and Ivan A. Bataev.

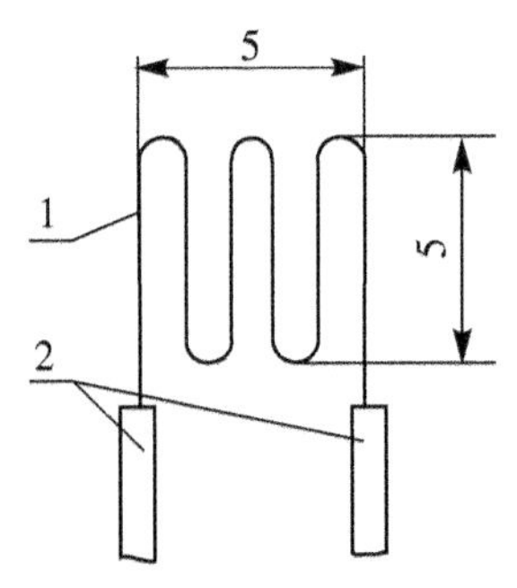

Figure B.1 Manganin gauge: 1, manganin and 2, contacts.

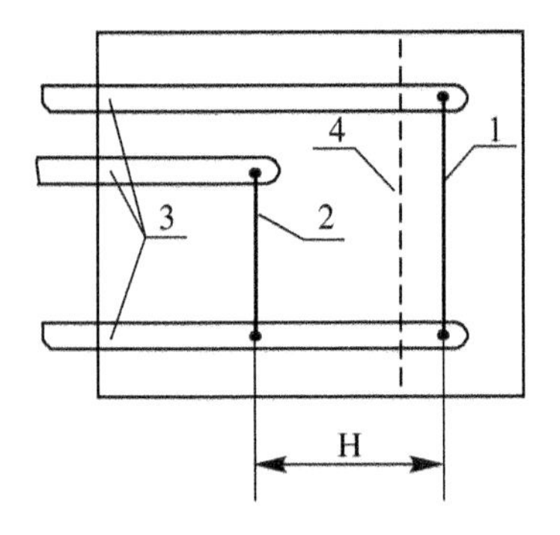

Figure B.2 Manganin gauge: 1, 2, manganin; 3, contacts; and 4, detonation front.

and in our experiments it was about 0.1 μs. In the experiments on measuring the pressure profiles at the explosive/ceramics interface, we used gauges in the shape of bands approximately 0.5 mm in width (Figure B.2), which were cut out of 20 μm foil. The length of the gauges was 30–40 mm. H is the distance between the centers of gauges. The use of such gauges made it possible to eliminate the front crash caused by the gradual loading of the area of the sensitive element. The only remaining crash was associated with the curvature of the detonation wave front and the possible nonparallelism of the gauges and the front.

All experiments using manganin gauges were conducted at the Institute of Problems of Chemical Physics of the Russian Academy of Sciences (IPCP RAS).

B.1.2 Optical Methods

Several optical methods were used to measure the parameters of the shock wave and detonation processes in a number of experiments. Optical measurements were performed in both streak and frame modes [14]. The ZHLV-2, SFR-2M, VFU, and SHIMADZU HPV1 cameras were used for measurements.

The optical methods were used to observe and measure the shape of the front of the detonation wave in various explosive charges, the behavior of ceramic samples under intense pulsed loading, etc. The experiments using optical methods were conducted at the Bauman Moscow State Technical University (MSTU), Lavrentyev Institute of Hydrodynamics, and Institute of Pulsed Power Science, Kumamoto University.

B.1.3 GAP Test

There are a large number of methods for studying the process of detonation transmission [3]. The design of most of the experiments described in this monograph corresponded to the simplest but fairly common method of investigating the critical conditions for the detonation initiation that involves loading of the analyzed passive explosive charge with a contact explosion of an active charge separated from it by an inert obstacle or air gap. In English literature, this method is called the GAP test; in Russian literature, it is often referred to as the barrier method. A typical experimental setup is shown in Figure B.3. In this case, the presence or absence of detonation transmission from the active to the passive charge is determined by the presence or absence of a fingerprint on the "witness" plate.

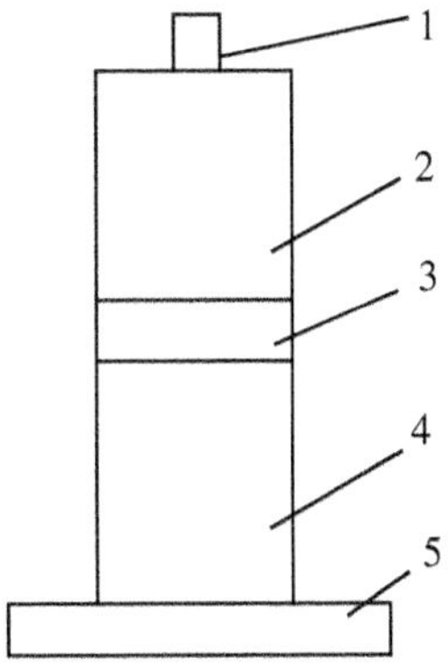

1
2
3
4
5

Figure B.3 The scheme of a GAP-test assembly: 1, detonator; 2, active high explosive (HE) charge; 3, barrier of inert material; 4, charge of the investigated HE; and 5, witness plate.

B.2 Simulation Software

B.2.1 The POTOK (ODVAX) Software Package

POTOK-ES was developed at the Institute of Applied Physics SB RAS [4] and is intended for the numerical computer simulation of nonstationary processes of explosion and impact in a two-dimensional formulation.

In this software, a numerical method of "individual particles" is used to integrate the equations of continuum mechanics [5]. This method belongs to a group of "particles in cells" techniques, in which a continuous medium is represented in the form of Lagrangian particles moving along a fixed Eulerian mesh. This method allows relatively simple calculation of flows with large deformations with a relatively large number of contact and free boundaries. The computational software implements the method of "individual particles" on the adaptive nonuniform rectangular Eulerian meshes.

B.2.2 The ANSYS AUTODYN Software Package

ANSYS AUTODYN [6] was developed by the American company Century Dynamics and is intended for numerical modeling of nonlinear dynamics of solids, liquids, gases, and their combinations in two- and three-dimensional formulations. This is a powerful tool for interdisciplinary computations in problems where loads change over a short period of time, as in the case of a high-velocity impact or explosion. ANSYS AUTODYN contains many solvers and computational methods that use

Lagrangian and Eulerian approaches to the flow of a continuous medium, an arbitrary Lagrangian–Euler formulation, as well as the method of smoothed particle hydrodynamics (SPH), which forms the foundation of the mesh-free solver. ANSYS AUTODYN is fully integrated into an easy-to-use graphical interface, the capabilities of which are complemented by the functions available in the ANSYS Workbench. These functions enable quick and easy preparation of a numerical model based on computer-aided design (CAD) geometry, generation of a mesh that is most suitable for analysis, and multiple computations using parameterization. Geometric dimensions, materials, and their physical and mechanical properties, as well as initial conditions (for example, the impact velocity), can be used as parameters.

References

1 Al'tshuler, L.V. (1965). Use of shock waves in high-pressure physics. *Soviet Physics Uspekhi* 8: 52–91.
2 Selivanov, V.V., Soloviev, V.S., and Sysoev, N.N. (1990). *Shock and Detonation Waves: Methods of Investigation*. Moscow: Moscow State University [in Russian].
3 Zhernokletov, M.V. (ed.) (2005). *Methods for Investigations of Materials Properties Under Intense Dynamic Loads Monograph*. Sarov: RFNC-VNIIEF [in Russian].
4 Agureikin, V.A., Vopilov, A.A., and Kulkov, O.N. (1987). *The Software Package POTOK-ES. User Manual*. Novosibirsk: Institute of Applied Physics [in Russian].
5 Agureikin, V.A. and Kryukov, B.P. (1986). The method of individual particles for the calculation of flows of multicomponent media with large deformations. *Numerical Methods of Continuum Mechanics* 17 (4): 17–31. [in Russian].
6 Century Dynamics (2005). *Autodyn. Explicit Software for Nonlinear Dynamics*. Theory manual.
7 Bridgman, P.W. (1911). The measurement of hydrostatic pressures up to 20,000 kilograms per square centimeter. *Proceedings of the American Academy of Arts and Sciences* 47: 321–343.
8 Fuller, P.J.A. and Price, J.H. (1962). Electrical conductivity of manganin and iron at high pressures. *Nature* 193 (4812): 262–263.
9 Bernstein, D. and Keough, D.D. (1964). Piezoresistivity of manganin. *Journal of Applied Physics* 35 (5): 1471–1474.
10 Khristoforov, B.D. (1971). Shock wave parameters associated with the explosion of a spherical charge in porous sodium chloride. *Combustion, Explosion and Shock Waves* 7: 507–511.

11 Kanel, G.I. (1974). *The Use of Manganin Gauges for Measuring the Shock Compression Pressure of Condensed Media.* VINITI, No. 477-74. [in Russian].
12 De Carli, P.S. (1976). Manganin stress gage calibration to 125 GPa. *Bulletin of the American Physical Society* 21 (11): 1286.
13 Espinosa, H.D. (1998). Recent developments in velocity and stress measurements applied to the dynamic characterization of brittle materials. *Mechanics of Materials* 29: 219–232.
14 Dubovik, A.S. (1984). *Photographic Recording of Fast Processes*, 1984. Moscow: Nauka [in Russian].

Index

Explosion Systems with Inert High-Modulus Components: Increasing the Efficiency of Blast Technologies and Their Applications, First Edition. Igor A. Balagansky, Anatoliy A. Bataev, and Ivan A. Bataev.